The Last OLD GROWTH Sawyer

World Saving Solutions by

KEVIN BRUCH

Printed and bound in the United States of America
First printing • ISBN # 978-1-954463-01-1

The Last
OLD GROWTH
Sawyer

World Saving Solutions by
KEVIN BRUCH

SCOTT PUBLISHING COMPANY
www.scottpublishingcompany.com
P.O. Box 9707 • Kalispell, MT 59904
Toll Free: 1-800-628-0212
Fax: 1-406-756-0098

DEDICATION

This book is dedicated to the family unit, The building block of society. More specifically to my wife Carrie, who propelled me forward during periods of frustration. When (although I'd never admit it) it seemed the only thing I'd discovered was a means of wasting my life in pursuit of the impossible. Thank you. I love you.

To my children Autumn and Michael and my bonus children Christopher and Jessica, along with my grandchildren Grayson, Eleah, and Dustin and those yet to be born. You've all been my inspiration and made life worth living. The same can be said of my extended bonus children, Kylie, Kristian and Cale. I love you all.

INTRODUCTION

The old Western Larch stump depicted on the cover has a story to tell. It's a tale that began long ago and involves how, why and when the Native Americans learned to live contently within the laws of nature. That benevolent coexistence of not so long ago had been born out of an all-consuming bleakness.

Several thousand years prior, an unknown ancient people had unwisely toyed with nature. This interaction eventually lead to a mass extinction event that wiped out several climate-influencing mega fauna species. In the absence of these all-important animals, the global climate collapsed.

Those ancient people achieved this unintended devastating outcome by foolishly forcing their 'will' on nature through the misuse of fire, by pursuing an ill-conceived fire hunting strategy.

Following the ecological/climate collapse, life descended into a living hell. Then, after thousands of years of an unrelenting harsh existence, the Native Americans began restoring nature with the 'wise and timely' use of fire. The climate stabilized.

......

That cover stump is now screaming at the arrogant 'political ruling class', who over a century ago began forcing its will on the arid western landscape - a landscape these politically powerful people knew almost nothing about.

Those far away politicians never sought to solve the complex mystery of how the Native Americans became the indispensable maintainers of nature. Instead, the political ruling class ordered them removed and immediately began installing European-style agriculture on the arid western landscape. Vast portions of that dry landscape would prove ill-suited for the task.

The homesteaders that replaced the Native Americans soon found themselves in a perpetual struggle for survival. Droughts, dust storms and forest fires are just a few of the challenges these families faced as they battled the often harsh and unforgiving landscape. This constant war of survival would all too frequently prove unwinnable.

My family was among the first to come west after the Civil War. They found an existence mired in tragedy and/or economic desolation. However, like their Native American predecessors, they would gain a hard-earned knowledge about their ecological surroundings. These trials and errors have been passed down through the generations. This unique family history has been the spark that has allowed this logger to make several unprecedented scientific breakthroughs.

This new knowledge base will allow us to harness the industrial world for the explicit purpose of stabilizing the climate by restoring what has been lost. This proposed radical paradigm shift will ensure that future generations have the opportunity and ability to live in harmony with nature.

These original discoveries are an indictment of the professional politician. This occupation, even with the immense resources of the federal government at its disposal, has thoroughly proven itself incapable of finding the solutions our complex society requires. The following pages will reveal how this archaic profession, through incompetence and/or deceit, created the climate change/oscillations issues we now face, which include the western mega fires.

Social media technology has rendered both the professional politician and mainstream media obsolete. Towards the end of Chapter 1, I unveil an envisioned 4th branch of government, which when implemented, will harness the collective empathy and intellectual might of the American people.

The 4th branch will save America from the ideological religionists, who've unwittingly become the agents for centralized authoritarians. These political tyrants have utilized the environmental movement as a means of destroying the unrealized original intent of America. We are now witnessing the final battle.

From founding, self-serving elected officials have usurped the immense power of government for themselves. 'We the People' must unite and save America, because only America as originally envisioned will provide the individual freedoms that real solutions demand. The 4th branch will provide that freedom, while finally fulfilling that long ago broken promise.

After implementation, we can begin the long process of restoring the ecological world.

We must hurry, because the point of no return draws near.

……

Listing the vast resource materials I've utilized here at the beginning would only serve as a distraction. However, there are two books that demand mentioning up front because without either, finding the true solution would have been impossible: 'Forgotten Fires' by Omer Stewart and 'Northwest Trees' by Stephen Arno. The world owes them both a deep debt of gratitude.

The same can be said for the many family members and loggers you will meet in the coming chapters.

FORWARD

Written by Rick Moore, Service Forester Montana DNRC

Gyppo logger. Merriam-Webster define it as a "small logging operator who usually works on a contract basis." They are independent, adaptable, mobile and proud. Margaret Elley Felt, wife and partner of a gyppo, famously stated he had a "virus of independence in his veins."

I have known many a gyppo logger. Every one of them coaxing their well-used equipment to get one more day. Yet most gyppos innovatively modify that equipment to grasp every measure of performance possible. Living on the financial edge, risk is a way of life.

Look up the 10 most dangerous occupations in America. "Logging workers" are at the top of that list. Logging can be physically demanding, frequently performed in remote places far from medical help. "Struck by an object" is the most common injury, and when you consider the array of objects, including trees, that are associated with logging, it is no wonder.

Today gyppos are more commonly known as "contract loggers." They are the backbone of the logging industry in northwest Montana. This in turn makes them a key element in the forest products structure of the same region. Names like Weyerhaeuser, Stoltze, Pyramid and Tricon rely on them to deliver. Further down the chain, Smartlam, Intermountain and Empire don't exist without them.

This is a story of one Montana gyppo family. It is written by a fifth-generation son who followed the logging lineage of his ancestors. Though the chainsaws have been replaced with feller-bunchers and in-woods processors, the work ethic and innovation remain. This logger and his brother are in the minority. They work the steep ground, the ground no one else can or will do, and they accomplish it with a cable logging system. Cable logging brings new meaning to the possibility of being "struck by an object."

Yet it is more than a story. It is a new product being introduced into the "marketplace of ideas."

It is a challenge to the status quo thinking of how we got to this point in the implementation of forest management in America.

Intwined between the compelling Bruch Family's struggles to survive in a dangerous and unpredictable occupation, is an equally compelling idea of how we got to this point in the climate change discussion. Relationships I had never connected involving indigenous peoples, fire, glaciers, trees and mastodons that shaped the land cover. A new paradigm in the marketplace. Fresh ideas for the discussion.

Arguably the ignorance, arrogance and failure of the political class to adapt to a changing world caused disorder in the relationships shaping the landscape. This moved the forest managers away from the potential solution to a problem they did not yet recognize. The recognition is now upon us, but the political structure to deal with it is not.

Delve into the story. Challenge your own view on the evolution of the forest, the foresters, and the politics. Open your mind to new ways of looking at change, to new ways of looking at the forest. When finished, you may never look at the forest the same way again.

Rick Moore
Service Forester, Montana DNRC

CHAPTER 1
A WORLD SAVING SOLUTION

Climate change is real and the direct result of mixing science with politics.

I never set out to discover why the climate has begun oscillating. This answer however, is just one of many that are now firmly in my possession, along with the required remedies. We must act quickly, because we are already witnessing the early symptoms of the reemerging Altithermal. If left unchecked, weather patterns will continue to degrade until the rains stop, at which point agriculture will no longer be able to feed the world's population. Famine will be followed by Armageddon, only to be followed by mass extinction and the final judgement.

I've quite by accident unlocked the mystery of us. It's a shocking story that will reveal how long ago, man began destroying nature with the misuse of fire. This includes the removal of an all-important climate-influencing forest structure that once dominated the Great Plains of North America.

The consequences of this activity ultimately led to both a mass extinction event and the collapse of the Pleistocene floral/faunal climate mechanism 11,600 years ago. We are living in the shattered ecological remains of that former Garden of Eden like system.

For several millennia following the collapse, Earth's climate spiraled out of control. The massive glacial ice sheets that once covered what today is Canada, slowly melted. By 8,000 years ago, the meltwater generated from the disappearing glaciers had caused the oceans to rise by 400 feet.

The next 3,000 years (8,000-5,000 years ago) is referred to, among other names, as the 'Altithermal'. It was a dry, bleak time. Water tables fell. Deserts formed and grew. Dust storms frequented the Great Plains. Populations plummeted.

It could have been game over, but then somewhere around 5,000 years ago, the Native Americans began developing restorative burning strategies/traditions. I've chosen this date because it was then that the bison, the sole mega fauna survivor, began to rebound. The bison however, reemerged at the end of the Altithermal as a much smaller animal. This 'stunting' provides evidence that following the ecological collapse, the quality and quantity of forage had drastically diminished.

The anthropogenic burning restored the grasslands and quite by accident, reinstalled the vital separation between two incompatibly-evolved tree types. This burning produced thick grass, which explains why the bison numbers suddenly began rising. The Native Americans, in order to protect and maintain this new found food source, established durable burning traditions.

During the Ice Ages, the division of forest types was the primary ingredient powering the Pleistocene climate-influencing mechanism. The most powerful forest of the Pleistocene epoch, in terms of its ability to influence the climate, was the aforementioned Great Plains forest. This claim is based on proxy evidence from Antarctica and Greenland ice core analysis and the timing of the mass extinction event, which suggests the Pleistocene climate collapsed immediately following its removal.

This photo clearly shows the transition. The Old Growth Larch tree was several centuries old, when it was cut down over a hundred years ago. In this particular area, the fire intolerant Grand Fir went on to replace that former park-like fire resistant forest structure.

While I'm not suggesting they tried, the Native American method of fire restoration was inadequate, in terms of its ability to reestablish the Great Plains forest. Any effort using fire alone would have been thwarted by both the incredibly strong prevailing winds and the aridness caused by the Rocky Mountain rain shadow. This vast area would therefore remain deforested.

You may be asking, if fire prevents a forest from establishing itself on the dry, windy Great Plains, why am I making the assertion that a forest not only resided there, but one with specific climate-influencing characteristics? The answer resides within the dietary requirements of the mastodon and the many polar opposite characteristics found within those two contrary tree types.

I've placed these trees into two groups and named them 'Inter-glacial' and 'Glacial'. Both tree types had slowly evolved over many 100,000-year-long glacial cycles. Each individual tree specie acquired its own unique characteristics tailored to fit its specific circumstance and needs, denoting long-term stable plant communities and lengthy periods of isolation. During the cycles, one forest type or the other would gain or lose ground depending on Earth's location within the glacial cycle. This ebbing occurred primarily west of the Rockies. Because of the Rocky Mountain rain shadow, the Great Plains forest likely remained steady. A fairly short and warm inter-glacial period marked the end of every glacial cycle.

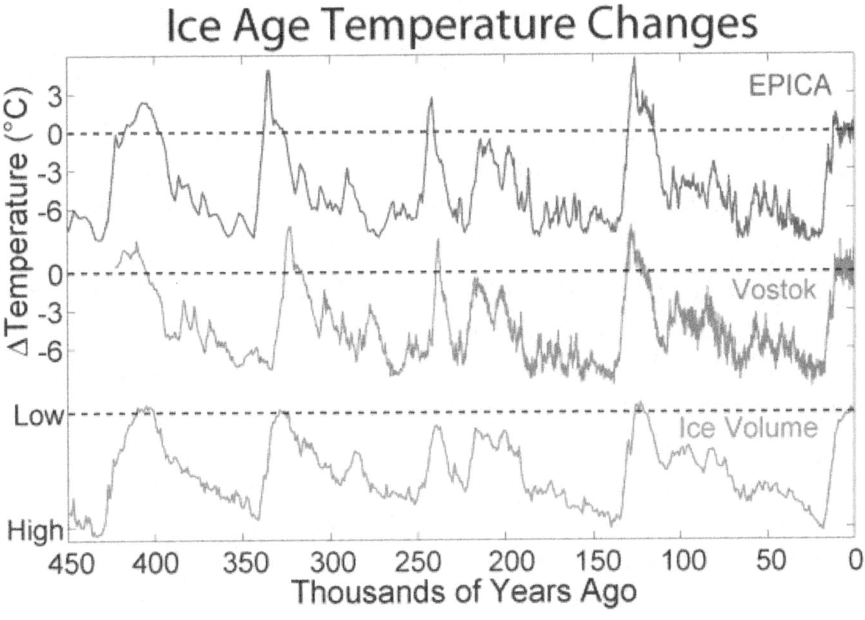

The many western tree species, regardless of type, are still functioning components of that broken, but repairable climate-stabilizing mechanism. It only lacks the functionality of one missing component: the mastodon. I've devised an industry that replicates the actions of the mastodon by mechanical proxy. When fully developed and implemented, this still-imaginary industry will supply vast amounts of wood fiber for a wide variety of products, which include huge quantities of structural lumber along with tremendous volumes of renewable electrical energy.

During the Pleistocene, the crucial tree type division had been maintained by the now-extinct mastodon. The mastodon removed the lower limbs from trees and while it may not sound like much, it was this action that inhibited fire-intolerant Glacial trees from infiltrating the park-like Inter-glacial forests that once occupied the Great Plains region.

Restoration of an American mastodon herd by Charles R. Knight.
While interesting, we must nurture the inter-glacial forest back into existence before pursuing thoughts of reintroducing extinct animals. To do so before hand will inhibit restoration.

The Inter-glacial forests are fire-adapted and possessed lush, grassy forest floors. It was here that the many mega grass-eating animals roamed. Most of these giant animals were completely dependent upon the Inter-glacial forest structure for survival.

Both trees types, depending on exposure, owe some of their evolutionary make up to either the mastodon or low intensity ground fires. The Inter-glacial trees adapted to these frequent disturbances with thick bark and the ability to heal wounds. The Glacial tree species however, were forced to reside within areas of elevated moisture. These very damp places protected most of the thin-barked Glacial tree species from the disturbances generated by either the mastodon or fire. This explains why many of the shallow-rooted Glacial trees never developed any fire-resistant traits or the ability to produce wound-healing pitch.

Man fatally interrupted the system with the misuse of fire. This eventually led to a mass extinction event. The mastodon was among the many mega faunal victims. In the absence of the maintenance previously provided by the mastodon, the contrary tree types merged.

The infiltration and overstocking of shade-tolerant Glacial trees growing under the high canopy of the Inter-glacial forests soon replaced the grass.

One of the evolved traits of the Glacial tree is resistance to losing its lower limbs. This eventually caused a problem for the Inter-glacial forests by creating a fire ladder. Fire could now climb from the forest floor up into the high and very flammable old growth canopy (Inter-glacial trees require direct sunlight and therefore shed their lower limbs when growing near other trees). Neither tree type could survive the ensuing fire storms. The subsequent mass deforestation shattered the Pleistocene climate-influencing mechanism.

The increased fire severity was in part caused by the desiccating effect of the Glacial tree intrusion, which inhibited the water storage abilities of the Inter-glacial forest structures. Inter-glacial forests protect and store water within the soil, whereas a Glacial forest removes it. The infiltrating Glacial trees, believing themselves to still be growing in a very wet aspect of the landscape, began removing the stored soil moisture from the Inter-glacial forest floor, discharging it into the atmosphere as water vapor. This drying out of the soils dramatically increased fire behavior.

The Pleistocene Inter-glacial forest, when functioning properly with its faunal maintenance components in place (the giant ground sloth was another Inter-glacial forest maintainer), was capable of transitioning arid aspects of the landscape into areas teeming with life. Unfortunately, the extinction event and the water-pumping abilities of the Glacial trees negated this terra-forming function and rendered the occupation of these sites untenable. Huge tracts of the western landscape became and remained deforested.

The later beneficial burning practices as performed by the Native Americans restored the critical tree type division where possible. These frequent fires forced the Glacial tree species (Inland Douglas Fir, Grand Fir and others) to reside within areas of elevated moisture. The Inter-glacial trees (Ponderosa Pine, Sequoia and others) were once again permitted to dominate the fire-prone aspects of the landscape. This restoration allowed the evolved characteristics of each tree type to be maximized, which soon began generating some positive climate values.

Although it was a very anemic representation of the lost Pleistocene system and covered only a small portion of the overall western landscape, it proved enough. Earth's climate began to stabilize. The rains returned. The Altithermal went dormant. The eastern forests were also undoubtedly a huge contributor towards climate stabilization.

Several millennia of this beneficial burning allowed these masters of their environment to transform their ecological surroundings, with the wise and timely use of fire. In so doing, the Native Americans became the indispensable maintainers of both fire-resistant forest structure and the global climate. Virtually everyone alive today owes their very existence to the former Native American burning traditions of North America.

When early European explorers arrived on this continent, they were awestruck by the beautiful landscapes. However, these vanity-driven agents from far away and Europe's competing centralized authorities rarely, if ever, attributed these breathtaking outcomes to the Native Americans. Instead, they arrogantly judged these people who were living within the laws of nature as an inferior race.

Despicable treatment ensued. In 1905, our overly-centralized U.S. federal government officially ended the former burning practices with a nationwide fire suppression policy. The two incompatible forest structures soon began merging.

All of our western forests have since transitioned into fire-intolerant entities and ceased to resemble those breathtaking descriptions of long ago. Worse, the climate-stabilizing values once maintained and generated by the Native American burning have also vanished.

The long-sleeping Altithermal is beginning to stir. If we don't act quickly, it will never sleep again.

......

Several decades ago, while removing some of the last remaining Native American-influenced forest structure as an old growth sawyer in Northwest Montana, I recognized the transition and the Catch-22 it had created. This insight was gleaned from my great-grandfather Ernie Everett, who recorded some of his memories just prior to his passing in 1986.

Long ago, he had also been a NW Montana old growth sawyer. For 121 years, multiple family members from each successive generation have worked the woods of Montana as loggers. He described an open park-like forest of his youth, with its grassy/wildflower covered forest floor. Great-grandfather Ernie followed up with a startling statement, a statement that went on to change the course of my life. Paraphrasing, he said, "The former frequent Native American burning had 'created' and 'maintained' that beautiful outcome."

His forest descriptions didn't match what I was seeing all those decades later. For the first time, I realized the forest structure had undergone a radical negative change. It had grown thick, the grass/wildflowers were frequently absent, and the forest was increasingly susceptible to canopy fires. I suddenly understood this wasn't natural. It was the end result of

This photo clearly shows the transition. The Old Growth Larch tree stump on the right was several centuries old, when it was cut down over a hundred years ago. In this particular area, the fire intolerant Grand Fir went on to replace that former park-like fire resistant forest structure.

This century old Grand Fir on the left (glacial) is located in the Swan valley. The dark area on its trunk, indicates that this tree is a rotting standing shell. It will break off and/or die in the coming years. Its apparent that the Grand Fir didn't evolve within areas possessing excessive snow, which is a common feature in the Swan valley. This tree species also didn't develop the ability to heal wounds. Any branch broken from a heavy snow load, will subsequently allow infection to enter the stem, which in turn causes it to rot from the inside out.

the federal government's 1905 fire suppression policy.

I became convinced the Native Americans had somehow, once upon a time, become the indispensable maintainers of fire-resistant forest structure. This confidence was elevated by the realization that I'd literally been surrounded by evidence supporting my great-grandfather Ernie's claim.

The most compelling clue was the ancient fire scars that marred many of the old growth trees. It was quite clear that these trees had survived

This dead Old Growth Western Larch (Inter-glacial) was recently cut down for safety reasons. It'd died, like millions of others, after having succumbed to decades of drought stress caused by the fire suppression induced glacial forest transition, which can be seen in the background. Those much younger trees are also dying. Many having grown on arid aspects of the landscape, where they are ill suited to thrive.

The ancient fire scar on the Larch stump presents itself as burning evidence of the former Native America wisdom.

countless burnings. Yet, not only had the fire-resistant trees healed after each event, but they had gone on to thrive for several centuries as a direct result of these trials by fire. Counting the growth rings revealed many of these old trees exceeded 700 years in age. The massive forest canopy fires that began in the 1960s were now frequently killing everything caught in those fiery paths, including these majestic sentinels of time.

I became perplexed. Why weren't our elected leaders discussing just how crucially important the anthropogenic burning had been during the presettlement era? After all, perhaps a viable solution was lurking within those murky unexplored depths.

......

Years later, I would discover there had been several others throughout the preceding decades who had arrived at a similar conclusion: John Wesley Powell, Omer Stewart, Elers Koch, Aldo Leopold and Stalker Leopold. All were highly acclaimed scientists.

Unfortunately, a political information firewall stood between them and the American people. This barrier was constructed for the sole purpose of protecting the occupation of 'professional politician'. These scientists had discovered an issue that conflicted with the false political narrative of their day, which had wrongly been characterizing the Native American as intellectually irrelevant.

Following the Civil War, the political class began aggressively inserting a European type culture onto the western landscape. They had assured the American people and the multitudes of 'European-descended' homesteaders heading into the western unknown that the Native Americans were being removed from the landscape in order to usher in a higher use of the land. It was a lie.

As stated, my family was among those early westward hopefuls. Each subsequent generation has unknowingly paid a steep price for the never ending political deceptions. Blissfully unaware of the many political manipulations, my ancestors pushed forward and soon developed a deep respect for the Native Americans. This appreciation was attained from many subsequent periods of trial and error, where life or death hung in the balance. These moments, which all too frequently proved fatal, gave a heightened perspective into just how tough it was to make a living from this often unforgiving landscape.

As the decades drifted by and more was learned about the parched landscape, individual scientists began coming forward to challenge the politically-constructed narrative. These scientists rightly concluded that the Native Americans had learned to properly maintain the forests and grasslands during the presettlement era. Several insisted that we must immediately return to the Native American way of frequently burning the landscape, undoubtedly sensing time was quickly running out.

This truth would have destroyed the credibility of the professional politician in the eyes of the American people by exposing the crime these charlatans had committed. They had blatantly misappropriated and misrepresented science, in order to force their own predetermined political outcomes onto the Native Americans, homesteaders and nature. Knowledge of this systemic deception would have enraged the American voter and likely put an end to this perpetual fraud by finally removing the professional politician from the decision-making process.

That didn't happen because it's the professional politician who controls the power and purse strings of the federal government/bureaucracy. The self-serving politician has created consortiums of like-minded travelers, which include many within academia. Together, they've usurped the will of the American people by harnessing the immense power of government for their own purposes. This has enabled them to both enrich and protect themselves from any inconvenient truths. This unconstitutional and therefore 'unlawful centralized authority' was/is further bolstered and maintained by a complicit mainstream media.

So instead of fixing the problem, each of these intellectually honest scientists found themselves either ignored and/or destroyed by the political ruling class. The last attempt was made in the early 1960s, at which point it had become too late.

The obscurity of the former Native American burning was perpetuated by the 'political ruling class', who for reasons that should now be obvious, didn't fund the necessary studies. As a result, a deeper understanding was never attained. This potential political 'wooden stake' slowly became forgotten.

There will be no one after me. The reason is disturbingly simple. The last of the Native American-influenced forest structures, along with the hard working, innovative 'gyppo' logging culture that sprouted up from the original homesteaders are now all but gone. The latter has been destroyed and removed from the landscape by a very reminiscent false political narrative campaign, which diminishes the credibility of a targeted group in order to bestow a second class citizen status onto its individual members.

This dishonest, but effective method is how the professional politicians deflect criticism away from themselves, while at the same time gain/maintain influence. The two party system allows one to play off the other.

……

All those decades ago, I had another moment of clarity in what has proven to be a disturbingly accurate prediction. The western forests were doomed, with a large percentage fated to die in unstoppable forest fires. Worse, this mass deforestation was going to occur regardless of whether we logged the forests or not. Something was needed to avert that dire future, but whatever that was, it did not yet exist.

Revealing my visions to others within my industry was met with near universal disbelief and quickly dismissed as absurdity. I would have reacted exactly the same way, had it not been for my great-grandfather's revelation and my own unique immersion within those storied old growth forests. Many other logging stories passed down by several more of my Montana ancestors were equally important, which collectively allowed me to comprehend what I was then seeing.

At the time, I had no clue about the associated climate implications. I only saw that we desperately needed to reintroduce fire back onto the landscape, but couldn't because the forest structure transition had crossed the point of no return back in the 1960s. I set forth to discover a means of restoration and as you now know, I've found so much more. I confidently proclaim possession of the necessary solutions, which when implemented, will slowly begin reversing our current apocalyptic trajectory.

……

All our contemporary climate change problems are the direct product of religious-like political ideologies, the expounders of which believe themselves proponents of a scientific movement pursuing a means of harnessing the industrial revolution for the betterment of all. In reality, they've only succeeded in centralizing the decision-making into the incompetent hands of a powerful few, who even with the best of intentions are incapable of leading us out of this mess. So before I begin expanding on the previously untold story, here is a partial list of initial and indispensable remedies that must immediately take place.

Again, we must hurry. The mega firestorms have already begun.

The Proposed 4th Branch of Government
(This is orignal and should not be confused with the current Wikipedia description)

1) We must abolish the political party system. This archaic governance method has proven itself not only incapable of solving the complex problems created by the Industrial Revolution, but is the very vehicle the political ideologues have used to drive us to this dire destination. The failed political party system is also responsible for keeping the original intent of America's founding from being realized.

2) We must unleash the unrealized potential that exists within social media technology. We can accomplish this by innovating and installing a system specifically designed to harness the collective empathy and intellectual might of the American people. This envisioned 4th branch of government will finally free us from the narrow-mindedness associated with far away centralized rule and the power parasites that are drawn to it like moths to a flame.

3) Remove all money from the electoral process. Those seeking office can utilize this new system to rise up through displayed wisdom earned from real life practical experiences and not an image constructed from false personas and empty promises.

Once elected, government compensation for the successful candidate must never exceed the median income of the private sector. Having our representatives live a similar reality to their fellow Americans will dissuade the self-serving individual, while simultaneously enticing the problem solvers to run for office. Once elected, these genuine people will have the immediate effect of improving the quality of decision making.

Lobbyists can appeal directly to the American people by accessing this new forum, thus removing the need for lobbyists to interact with our elected representatives.

This new way forward will provide immeasurable benefits for our elected representatives, as it will give them real time feedback and suggestions from those they represent. It will also create a direct conduit into the public consciousness for those few individuals who have spent their lives pursuing solutions to difficult and complex problems. This still nonexistent system will remove the roadblocks innovators have persistently experienced while attempting to reveal breakthrough ideas, especially those ideas that threaten the powerful.

4) Lawyers must not be allowed to serve in Congress. This occupation has its own branch of government and therefore must confine itself within the judicial branch. This will remove the conflicts of interest that have diminished the credibility of the justice system. It will also restore the vital separation of power between the branches of government, as was intended.

The justice system has become horribly unjust, which is evidenced by the different outcomes individuals are receiving for the same crime. It's increasingly apparent that verdicts are tied to an individual's elevated societal standing, net worth or lack thereof.

Therefore, in order to restore public confidence, the justice system must be demonetized. Justice must never be tethered to someone's inability to pay a fine, nor simply overlooked because an individual has enough money to settle 'out of court'.

Predatory litigation must also come to an end. Too many lawyers, who innovate nothing, have gotten extremely wealthy by attacking those who do. Removing the wealth incentive from lawyers will benefit everyone and go a long way towards mitigating the corruption that has currently infected the judiciary.

5) The central aspect of this proposed 4th branch of government must focus on harnessing specific individual practical knowledge and insight residing within each occupation. This potential governmental oversight, performed by a voluntary collection of otherwise unassociated individuals, should be utilized to educate their fellow Americans on unfamiliar topics and ensure our Representatives remain firmly tethered to reality.

American history would have been very different had this system been added to the U.S. Constitution at its inception. Slavery would have been immediately abolished. There would have been no Civil War and no genocide committed against the Native Americans.

There's one more possibility.

If this still-imaginary 'inclusive, decentralized, unpaid branch of government' had been added to the U.S. Constitution, it's quite possible that our brothers and sisters around the world would have adopted this system for their own and in so doing, perhaps cast off their own tyrannical oppressors. Maybe it would have even ushered in 'Peace on Earth'...

……

We all know this didn't happen, because unlike us, our American predecessors didn't possess the required social media technology. This very real remedy is available to us. We must seize the moment.

Contact your elected representatives and demand the U.S. Constitution be amended to include the 4th branch of government. The development of this system must be in full view and with the direct participation of the American people.

CHAPTER 2
FEAR AND FRUSTRATION

Long ago, while still in high school, a set of identical twins named Jody and Judy were friends with this old growth sawyer's son. We were young and idealistic with dreams of making the world a better place. Our conversations often centered around my family's long involvement with removing some of the last remaining old growth forests within the United States. One question always seemed to arise. How can the western forests be saved?

We were all of the same mindset. I despised European style tree farming (clearcutting) as much as they did. I'd reached that conclusion a couple of years prior, while working in the woods with Dad, during my first full summer of logging in 1979.

The moment of clarity arrived sometime during that inaugural summer. I'd asked Dad an innocent question, "How long would it take to regrow the majestic Old Trees". His response nearly ripped my heart out when he said, "They will never again be allowed to grow so old or large." He went on to explain, "We are removing these native forests to make room for European style tree farming, which is capable of growing wood fiber at a remarkably fast pace when compared to that of nature."

I balked at his explanation so he continued, "The American timber industry is actually saving other old growth forests somewhere else on the planet by satisfying modern society's ever increasing demand for wood products. And we are doing it on a relatively small percentage of the overall global forested landscape."

He finished with a scenario, "If this scientific forestry method were suddenly stopped, the results would be catastrophic because the exploding world population would fairly quickly destroy all the remaining natural forests on Earth."

His logic was unassailable, but it didn't make me feel any better and I yearned for a forestry method that allowed the grand old trees to live out their entire lifespan. At the time, this very poor high school student hadn't a clue about how to go about accomplishing the seemingly impossible feat of 'having our cake and eating it too.' So I shared my daydreams with the only two like-minded people I knew.

My Boot Camp photo

After high school the sisters went onto college, while I joined the Marine Corps. Four years later, I returned home and rejoined Dad's logging company, where I began killing hundreds of old growth trees each week. My distaste for removing these sentinels of time was palatable, but I consoled myself with a couple of thoughts. First, if I didn't kill them, someone else most certainly would, and second was my strong held belief that the best way of finding an alternative forestry method was from 'within' these fleeting forests.

This is my cousin Greg Dennison (Aunt Virginia's son). Greg and I were sawing partners in the latter half of the 1980's. Working near, but independent from one another, we fell ten's of thousands of Old Growth trees. In 1986, Greg dropped this 6' Western Larch giant (notice the thick heat shielding bark) in the Ashley creek drainage. This ancient Larch was by far the largest either of us encountered.
Today, Greg and his two sons (Curt and Larrik) are among the very few loggers still working the forests of Northwest Montana. Their efforts focused on forest stand improvements (fire intensity mitigation). Most old photos, unless otherwise indicated are provided by my Sister Rene' Bruch-Bishop. Who is our family's photo curator.

I was however, still at a complete loss on how to proceed, because of a seemingly insurmountable obstacle. I saw myself as under-educated and came to the realization that the sheer volume of knowledge it would require for such an endeavor wasn't available to me if I remained working in woods. I soon lamented this Catch-22 to Dad, who by now was well aware of my determination to find a better way. And although he was still very much a proponent of European style tree farming, he also adamantly believed loggers were innovators first and harvesters second.

He then opened my eyes to an almost boundless education opportunity swirling around us all. He said, "Everyone you meet will know something you don't. You simply need to observe the world around you and ask questions." This sound advice eventually led me to the solution.

Pursuing his advice, I began mining the minds of anyone and everyone: loggers, foresters, mechanics, welders, machinists, the list goes on and on. In a very real sense, everyone I met unwittingly became my professor.

My pursuit of an alternative forestry method however, abruptly came to an end or so it suddenly seemed. Pat Crawford, a logger from Wisconsin, had beat me to it. He'd shattered existing logging paradigms with the introduction of his remarkably innovative feller buncher (Timbco). This machine, along with mechanized log processors and grapple skidders ushered in a new era of mechanical harvesting, which drastically reduced the tree size requirements it took to maintain profitability. By 1990, this giant leap forward rendered previous 'conventional' logging methods in Northwest Montana all but obsolete.

1993 Timbco 425 Feller Buncher, photo by Forestry Equipment Sales

Dad was summarily given a choice by Plum Creek, the very large private timber company for which he contract logged. They told him he must either buy into the new technology or transition into line logging (yarderlogging). The choice seemed obvious to me, as I quickly saw myself in something shiny, new and on the cutting edge of technology.

Dad did not. He balked at the very high price of transitioning. The steep cost however, was less of a concern to him than were the young college students protesting in defiance of European style tree farming. They were chanting slogans such as 'Stop the Clearcutting' and 'Save the Old Growth'. I told him about my friends and reassured him that they weren't interested in destroying the timber industry and only wanted to hold us to a higher standard.

Dad wasn't so sure and pointed out that the court system was now involved and had already begun stopping federal timber sales. Undeterred, I once again attempted to bolster his confidence by saying, "The environmentalists were only appealing these timber sales as a practical means of showing us just how serious they wanted change." I furthered my impassioned appeal by claiming, "The environmentalists still hadn't realized that this new logging tech had made their argument just as obsolete as conventional logging had suddenly become." Unconvinced, he extended an olive branch, "I'll make you a deal, if we don't like line logging and the environmental movement proves genuine, we'll switch over to mechanical logging."

I was angry at him for not trusting me or the environmentalists (I would be proven oh, so horribly wrong), but I reluctantly agreed to be the sawyer for this new venture on a couple of conditions. I didn't have to be the leader or part of the skidding crew. What a self-centered ass I was.

……

Dad soon bought an old line machine, hired a crew and moved it to the first logging unit, all without my involvement. The yarder was a converted Linkbelt crane, manufactured in 1956. My first glimpse of this relict deeply offended my vanity. The paint that wasn't missing was faded, which as it turned out, proved a perfect match for its overly-experienced engines (one in the carrier and one in the crane). Opening a side door revealed rows of very large grease covered open gears, all spinning for some mysterious and seemingly menacing purpose. My heart sank. It all just felt like one giant and horrible step backwards.

Dad's 'Yarder'.

Plum Creek had given Dad a grace period, in order to allow us time to convert into the unfamiliar logging method. Dad remained with the conventional side and let Mark Marsh run the line side crew. The two crews were working different drainages far removed from one another, and therefore didn't have contact during the work day.

Mark had worked with Dad for a couple of years and was the obvious choice for leader, both because he actually had line logging experience, but perhaps more importantly, he was held in high regard by everyone on the newly-formed crew. The entire crew of five had logged before, but excluding Mark, all were as new grass when it came to line logging.

Line logging exists for the sole purpose of extracting timber from steep slopes that exceed the work limitations of mechanical logging. Mech logging is like power boating: you push the throttle forward and drive straight towards your destination. On the other hand, line logging is more like sailing: all day and every day you must 'think' the wood up the hill, as the crew journeys its way across very turbulent sea of potential problems.

Yarderlogging is an intense team game, where everyone must quickly do their part in order to achieve forward momentum. The crew only goes as fast as its slowest member.

As a sawyer, I was on the same hillside but out of eyesight from the skidding crew. I could however monitor the 'game' on the two-way radios we all wore. As the summer drifted on, it became increasingly apparent that our newly ordained yarder operator had become frightened of his new position.

His fear wasn't unjustified. The crane was held precariously upright by a cable (guyline). The guyline originates from the guyline drum and runs up the boom, through a large steel pulley held by a shackle, and back to a stump of unknown structural integrity. The cable is purposely left a little loose, so the weight of the machine can be brought to bear and therefore reduce some unnecessary strain on the stump.

This particular yarder had two other steel cables, both running downslope. The 'skyline' allows a carriage to remain elevated as it travels down and back up with a 'turn' of logs. The other cable is the 'mainline', which runs through and locks into the carriage as it pulls the logs.

The aforementioned slack in the guyline causes the yarder to heel unnaturally forward when the skyline is lifted, giving the operator the uncomfortable sensation that the machine is tipping over. The chocker setter (also known as the 'hooker') works downhill from the crane and is usually out of eyesight while 'hooking' logs. Its therefore imperative that

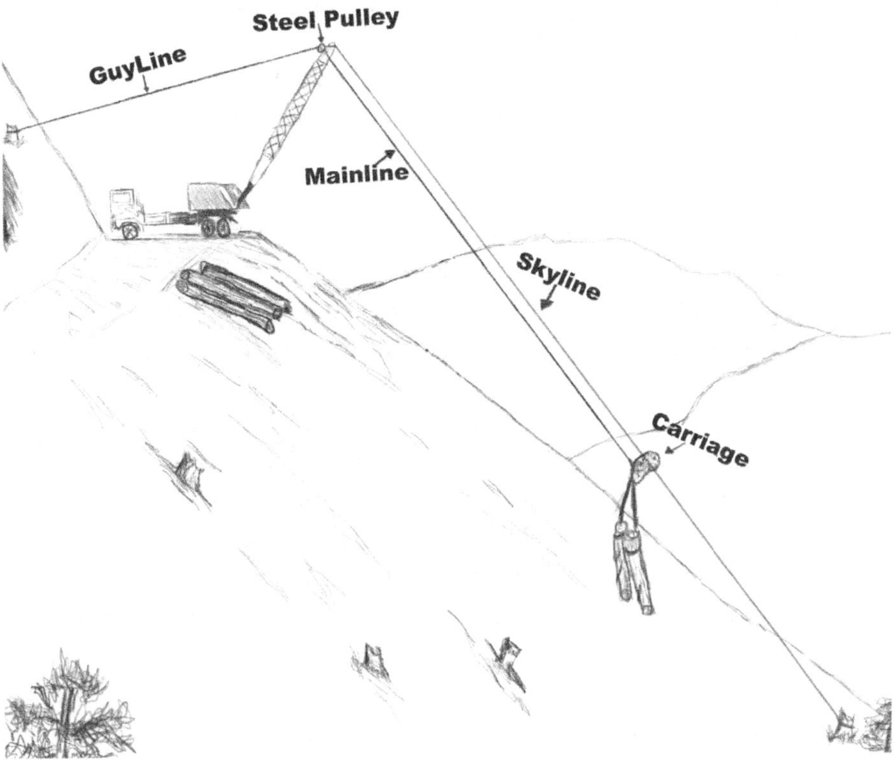

Example of a system utilized on dad's yarder, illustration by Michael Bruch

the yarder operator have an unwavering trust that the hooker will only 'hook' pullable weight. This trust is the most indispensable production and safety asset a yarder crew possesses.

When the hooker says 'take it', the yarder operator's butt instinctively grabs hold of the seat cushion as adrenaline-fueled anxiety permeates the mind in an attempt to feel everything, all at once--- Are the logs moving? You glance at the main drum--- yes--- feels heavy though--- did the guy stump just give a little--- maybe…no, its only my imagination--- You 'feel' the mainline lock into the carriage and know the 'turn' is now in the center of the corridor and it's time to give it full throttle…you glance over at the main drum and are filled with a sense of satisfaction as the cable begins building--- BAM, the machine shutters, oh crap just hit a stump… thank god the 'turn' jumped over--- BAM, the logs hit another stump and lock in behind it--- the guy stump, which holds the crane upright, pulls out in an instant. The yarder lurches forward, threating to break over of the center of gravity, which would send you and the machine down the hillside. Your only thought is with those who've already taken that unplanned journey and found the chances for survival very poor. The machine hesitates, teetering as if unsure of the appropriate fate it should bestow upon you. You plead your case to an unseen entity, seeking an intervention on your behalf--- Your prayer is suddenly answered and just as quickly as it began, the prospects for life are once again restored, as the yarder slams backward with a jolt.

Your nerves are frayed, but in as calm of a voice as you can muster, you push the transmitter button and inform the chocker setter that you've hit a stump 'down there' and 'pulled' the guy stump. The hooker responds with a question--- Is there another stump behind you? You look--- there is--- you feel a quivering knot in the pit of your stomach--- you want to go home but you don't reveal this truth. Instead, you say, 'Give me a few minutes, I'll notch it and put the guy line around the new stump".

The hooker informs you that he's already walking up the hill with a chainsaw in order to cut off the stump that had stopped the 'turn'. Again, forward momentum can only be achieved when everyone is doing their part.

Every yarder operator feels the fear. The good ones don't let themselves become ruled by this very natural emotion, because it blinds you to the reality of the moment and inhibits your ability to 'feel' what's happening while bringing in a turn. As dangerous as this job is, its actually statistically far safer, when compared to chocker setting or felling timber (cutting timber is the most dangerous job in the world).

Sometime in early summer, Dad's newly-minted yarder operator had pulled a guy stump. The experience had unnerved him and from that day forward, this truth was evidenced by his overly cautious operation of the yarder and subsequent radio chatter. His transmissions were usually one of the following: "It's too heavy" or "I stopped because it felt like I hit a stump" or "You'll have to cut it in half, because it's too big." Production began falling well below the break even mark.

Mark Marsh was the lead hooker and becoming increasingly frustrated with the low production numbers, which were a direct result of the lack of trust the operator was showing him. Friction between Mark and the operator continued building as the summer wore on, which could be heard in the often heated conversational tones emitting from the radio. As serious as the situation was becoming, some of Mark's responses to the inevitable 'question' were quite comical...

Example 1) Operator: "What am I stuck on?" Mark quips, "A

Huckleberry bush, GOOOOOO."

Example 2) Operator: "I think I hit a stump." Mark responds, "You did, it was rotten and I kicked it out of the way, GOOOOOO."

Example 3) My all time favorite was when Mark asked the operator, "Do you have a stick?" Confused by such a strange question, the operator hesitated with his reply--- "Nooo, why?"--- "Well, you'd better find one, because I got one and I'm coming up to kick your ass." Of course this wasn't true, but it was Mark's way of expressing his absolute frustration with the operator's lackluster performance. And as it would turn out, we would all soon learn a valuable lesson. Frustration can be just as deadly as fear and for the same reason, because it takes your focus off the task at hand.

......

The events surrounding the horrific logging accident that occurred on 10 September 1990 remain vividly etched in my mind. It was a beautiful fall day...

I'd stopped cutting timber a few days earlier because of the growing threat of having the unskidded logs becoming buried by the rapidly approaching winter snow. Instead, and for the purpose of bolstering production, I'd been 'high banking' (falling trees that would reach the road). After felling a few trees, I'd use a rubber-tired skidder to pull the trees to a centralized 'decking' area, where I bucked them into logs. In order to save time, I transported the chainsaw inside the cab with me, as

This second generation rubber tired skidder (Caterpillar 518) was the very machine I was operating, when the Accident occurred.

I alternated between felling and bucking. Big mistake…

I'd just pushed some logs off the side of the road and was backing up, when the chainsaw slipped and became stuck under the brake peddle. I frantically tried to free it, because the rearward moving skidder was rapidly approaching the edge of the road. I knew if the back tires went over the bank, the incredibly steep hill would cause the skidder to flip over backwards, with nothing to keep it from somersaulting down the hill until it reached a very distant bottom. Adrenaline-fueled fear began welling up within me as I mentally began preparing myself for this approaching prospect.

I was still tugging at the saw when a tire slipped over the edge. It was at this last moment that the saw came free from the brake. I slammed down hard on the brake peddle and the machine stopped. With shaking knees, I drove back onto the road. Someone in the distance was running towards me, having undoubtedly seen everything that had just transpired.

I sat there embarrassed, cursing my stupidity and vowing to never again bring a saw into the cab with me. I glanced over at the approaching runner. It was Mark and he was now close enough for me to see the look of abject terror on his face, sending chills down the length of my spine.

He stopped, out of breath and just long enough to say, "It's your Dad… he's hurt real bad…I'm going to the pickup… to use the long range radio… and call ALERT" (rescue helicopter). He turned and ran. I parked the skidder, got off and ran as fast as I could in the opposite direction towards the yarder.

Dad had came up for the explicit purpose of running the yarder. He'd hoped to get us back on track because our transitioning 'grace' period was quickly coming to an end. He'd become increasingly frustrated with the crew's lackluster performance, along with the abysmal fact that we'd grown no more proficient at this most extreme type of logging than when we'd started.

As I approached the machine I slowed to a walk. It was still upright, but eerily silent. I walked around behind the sideways-pointed yarder to gain access to the operator's compartment. Loose guyline cable was laying everywhere. Looking up at the top of the boom I noticed the large steel pulley, in which this cable was supposed to be running through, was gone…

I climbed an outrigger and stepped onto the deck of the carrier. I hesitated, frightened of the mysterious contents that cab now held. A moment later, I steeled my nerves enough to climb the steps into the cab.

I peered in, and my heart suddenly froze with the most gruesome sight my eyes have ever beheld.

On the way up, I'd solved the mystery of the missing steel pulley. It was laying just in front of the yarder, its placement revealing the cause of the accident. The shackle, which held the 70-pound steel pulley in place, had broken under extreme load, causing the pulley to slingshot down into the cab, entering through the top edge of the window. It had hit Dad on the front portion of his hardhat, which drove his head downward and through a lever.

Miraculously, he wasn't dead, but sitting upright when I'd climbed into the cab, having already pulled himself free of the lever. Brain matter was coming out of the front left portion of his shattered skull and his eyes were hanging out. His left shoulder was obviously broken and hanging unnaturally lower than his right.

I asked him what had happened. Blurrily, as if waking from a deep sleep, he stated he didn't know. Jim Pound, who was one of Dad's old sawing partners from the 60s, was there that day and now climbed up into the cab with us. I was grateful for the company. Together we tried to keep him awake, fearing unconsciousness meant death.

We asked him how old he was, and he slurred "51", which was correct. Each time he attempted to go unconscious I poked him in his broken shoulder to wake him back up. Every few minutes, we'd again ask him his age. Strangely, the number kept consecutively getting lower, in what seemed like a slow and steady countdown to the afterlife.

Unknown to Jim and myself was the struggle Mark was having contacting the rescue helicopter. He'd gotten on the radio, but discovered he couldn't hit the 'repeater', which was located on a mountaintop dozens of miles north of our logging unit (Big Hole drainage of the Thompson River valley). Luck was on our side however, because Ken Howard was hauling a load of logs thousands of feet lower in elevation, on the main 'river road' and had a better angle at the 'repeater'. Ken stopped, called ALERT and stayed there as a liaison while the helicopter was in transit. Another member of the crew got on the dozer and made a suitable landing pad on a rare flat spot, about 100 yards away from the yarder.

My first view of the chopper filled me with unbridled elation, but this euphoria quickly turned into anticipated horror. The helicopter was flying low and was about to pass in front of the yarder. The skyline was still up and high in the air and I hadn't a clue on how to go about lowering it. I therefore could only watch and cringe at the catastrophe getting ready to unfold before me. At the very last moment, the pilot either saw it or perhaps by mere chance, lifted his machine upward and over the skyline, just missing it.

ALERT landed and the logging crew soon emerged carrying a gurney, accompanied by the flight nurse. Dad was still conscious and disturbingly claiming the age of 5, as we slowly lowered him from the cab and placed him on the awaiting receptacle. We loaded him on the helicopter and backed away. As the chopper lifted and began flying towards Kalispell. I knew it was the last time I'd ever see him…

Mark and I got into my pickup and began the two and a half hour journey towards the hospital. Adrenalin-fueled worry was causing me to not realize that I was driving way too fast down this narrow, steep terrain logging road. In a calming voice Mark said, "He's in good hands." Grasping his intended meaning, I took a deep breath and eased up on the throttle. The situation was indeed out of our hands. I slowed down.

A half mile later while going around a wide, hard right corner, I was suddenly startled by an ambulance with lights flashing. I slammed on the brakes, but it was too late. I hit it in the side, just behind the driver's door. It would have been much worse if I'd been driving at my previous pace, but as it turned out, no one was hurt and the ambulance received very minor damage.

The ambulance had been dispatched from the small town of Thompson Falls and hadn't heard on the radio that this emergency vehicle was no longer needed. She now informed me that I couldn't move my pickup and we must both wait for the Highway Patrol. I had no intention of staying and I had a way out. The ambulance driver had been hugging the inside bank while going around this very wide corner. As a result, there was enough room to get a loaded logging truck around the crash site. The truck was heading to Kalispell, so I left Mark with my pickup and hitched a ride.

Hours later, a Highway Patrol Officer called me on the phone located in the hospital waiting room. He informed me that I would be receiving two tickets: one for leaving the scene of the accident and another for reckless driving. I thanked him and hung up. It seemed so trivial, in comparison to the news Mom had given me when I first arrived at the hospital.

Miraculously, Dad was still alive, but had been placed in a medically induced coma. A brain surgeon had told her there was nothing he could do, and he was more than a little surprised that he was still alive. He also said that if by some small chance he did survive, he would be a very different person.

Sometime shortly after receiving this news, a truck driver showed up to give his sympathies. As he was leaving, he turned and reminded Mom of a load ticket located in Dad's wallet, undoubtedly fearing it would become lost in the confusion and he wouldn't get paid. Yes, it was distasteful, but for me it was a reminder that the accident had been entirely my fault.

The trucker's concern for Dad had been genuine. Acquiring money however, for almost all of us, has become the primary driving force within our society. Money is not only the singular ingredient that enables the modern citizen to avoid a harsh poverty stricken reality, but with enough, one can actually manufacture a fantasy world detached from the merit-based laws of nature.

I'd been guilty of putting my own monetary interests above everything else. I'd clearly seen the situation deteriorating to the breaking point, yet I did nothing. The reason was simple: I made more money as a sawyer.

The consequences of my self-centered attitude had caused Dad, out of panic-driven desperation, to come up and run the yarder. His mind was mired with financial frustrations when he should have been focusing on running a very dangerous machine - a machine he had almost no experience operating. He therefore had never pulled a stump and lacked the appropriate level of caution. The last thing he'd said to Mark was "wood me up" and when the inevitably heavy 'turn' hit a stump, instead of pulling the guy stump, a shackle broke.

I should have volunteered to operate the yarder months earlier. Dad had asked me several times that summer to do just that, but I refused. My family would forever be haunted by the consequences of that day, which were a direct result of my own selfish monetary motivations. I vowed to never again place money in front of doing what's right.

......

Dad would languish in a coma for about a month. Then suddenly one day, he awoke strangely alert. He went on to make a full recovery, his intelligence still very much intact. Had he changed? This topic has been a source of debate within my family ever since. Some say yes, most say no. I contend he did, but the mild transition occurred after another unpleasant experience.

It was the beginning of the 'War on the West'. Lawyers and politicians would ultimately go on to destroy the western timber industry along with its innovative culture, for nothing more than their own political gain and wealth. They were cashing in on the emotional frustrations being generated by the environmental movement.

Two things were painfully clear: Americans were fed up with altering the natural world, and the logger had become the focal point of that anger. We were indeed the ones removing the forest, but loggers were merely following the instructions of university-educated foresters. The foresters themselves however, were equally powerless to change how the forest was managed.

Ironically, those decisions had been and were still being made by politicians and their political appointees, starting way back in 1905. The primary concern of the professional politician, above all others, is to protect their own occupation. When cornered, they simply assign blame away from themselves.

Teddy Roosevelt was the first progressive president. He sought to harness science and the Industrial Revolution for the betterment of all (cough). He helped usher in the Forest Service in 1905 and appointed Gifford Pinchot as its first chief.

European style tree farming at the time, was considered the height of scientific advancement. This economic-driven method removes the existing forest and replaces it with a monoculture manmade version. Once marketable sizes are attained, all the trees are cut down and the process repeated.

Gifford adopted this method as a one size fits all, for all of America's forestry needs. A fire suppression policy was put in place for the purpose of protecting the product value of the forest. To be fair, there weren't any other options, and excluding the one you're currently learning about, there still isn't.

The legendary naturalist John Muir supported the scientific tree farming option, but Muir's primary focus was on setting aside a few special places as sanctuaries, where nature could flourish in the absence of man. He was instrumental in turning Yosemite into a California State Park in 1890.

Muir didn't trust State politicians. This mistrust was at the forefront of his mind when he went camping with President Roosevelt in 1903. Muir encouraged the Chief Executive to make Yosemite into a National Park, with federal protections. Teddy agreed and signed it into law in

1906.

It's important to make the distinction between the three (Roosevelt, Pinchot and Muir). All three wrongfully believed the Native Americans were intellectually irrelevant. Roosevelt and Pinchot's viewpoints were mired in racism. Muir on the other hand, had come to his inaccurate conclusion by an honest misinterpretation of evidence. This moment occurred when he first entered what would later become Yosemite.

In the spring of 1868, after walking across the treeless, wildflower strewn Central Valley, Muir entered a forest. He witnessed evidence of massive burn injuries on many of the ancient Sequoias. The following decades saw his observations galvanize into a theory. He believed the Native Americans had long ago removed the forest from the Central Valley and were still in the process of removing what remained. Although Muir saw the importance that fire played in maintaining the health and dominance of the Sequoias, he also believed lightning was a sufficient source. Muir therefore asked the government to remove all 'Indians' from within Park boundaries.

Today, these protected areas lay in ecological ruin. It never dawned on him that the modern Native Americans had become the indispensable maintainers of nature and those wildflower fields were a product of a durable and beneficial burning tradition. Nor did he comprehend that the true guilt resided with an unknown group or groups of ancient people, who had perpetrated the original catastrophic deforestation event, which had occurred at the end of the Pleistocene and caused the collapse of the global climate.

The intellectual damage had been done. Academia went forward studying either the Muir doctrine or the Pinchot doctrine, each equally flawed. The former wrongly believed nature would revert to its true self in the absence of man's influence, while the latter sought its industrialization. These two polar opposite academic directions were destined to pull apart, especially since the power of the federal government was unequally skewed on the side of industry.

Muir had always feared the politicians would one day sell out and destroy the small fragments of nature he'd worked so hard to preserve. So in an effort to have some political leverage, Muir created and became the first president of the Sierra Club, which quickly gained political power. The following decades saw constant conflict between the two contradictory viewpoints.

Ironically, foresters following the Pinchot doctrine were the first to recognize that the Native Americans had been the maintainers of fire-resistant forest structure and grasslands. Several came forward and were immediately ignored.

Those following the Muir doctrine became aware of the warning, but simply edited the warning to fit their own well-established political narrative. In the 1960s, they helped institute a 'let it burn' policy within National Park boundaries, maintaining their original flawed idea that man must remain excluded from the process.

After the court system got involved in the late 1980s, the balance of power tipped towards the environmentalists. Strangely, the modern environmental movement suddenly morphed into an amalgamation of the two incompatible ideologies: Progressivism (everyone should have a house) and Naturalism (no trees should be cut down). This was the dawning of our modern era, with its frequent displays of astounding political hypocrisy.

My family suddenly found itself marked for destruction, forced to pay the price for something completely out of our control and not of our doing.

As stated earlier, a few years before this nightmare began, I'd come to the realization that the Native Americans had indeed become indispensable to the fire-resistant forest structure. In the absence of the former frequent burning, the western forests had become fire intolerant. I felt I was onto something original. I was completely unaware of the academic conflict or those few foresters who had come to the same conclusion a few decades earlier. I began daydreaming of a way to replicate the Native American success through mechanical proxy, my vanity consumed by visions of wealth and fame.

Then my family's new reality came crashing down. My self-centered fantasies were forgotten in the aftermath of Dad's accident. I was determined not to fail him again.

......

Sometime in 1990, millions of 'mostly' lodgepole pine trees located in the Challenge Creek drainage near Marias Pass blew over. The lodgepole weren't old growth trees (lodgepole have a short lifespan of 80-100 years), having replenished the area after the 1929 fire.

Half a dozen logging companies descended on the area in early summer of 1992. Within a month or two, Dad's company was the last one remaining. The rest had quit, because of a rash of sawyer accidents. The pixy stick like, weaved-together blowdown was a constant lesson in physics, with a grade of pass or fail immediately administered. Sometimes just touching a tree with your chainsaw would instantly release the pent up load and perhaps shoot the tree upward ten feet or sideways with tremendous force. One logging company experienced a broken leg. Other companies suffered a wide variety of similar impact injuries.

A sawyer working for Dad had a fully revved chainsaw pushed back into his thigh. Chainsaws don't cut, they rip, and the femur was exposed in what resembled a very large shark bite. This era remains the most consistently dangerous period of my life.

As the other logging companies departed, one by one, we stayed and inherited the abandoned units. Dad's accident had left him financially weakened and therefore not in a position to say no. Remarkably, we managed to stay accident free for over a year.

Then something happened, At the time, we were only a few days away from finishing. The Challenge Creek blowdown timber sale had been appealed in federal court by environmental groups. They'd utilized the 'Endangered Species Act', citing the endangered bull trout as the reason for the appeal. They feared sediment from the salvage logging would enter the stream and clog spawning beds.

This threat was real. So in order to prevent silt entry, wide setbacks were put in place, where logging wasn't permitted near streams. Therefore, the blowdown trees near streams would remain in place. This common sense decision allowed the salvage sale to proceed. It would prove to be one of the last.

During the harvest phase, a van load of very angry environmentalists showed up and if glares were lasers, we loggers would have instantly burst into flames. Their behavior wasn't an accident. The movie 'FernGully: The Last Rainforest' had been released a few months earlier. It depicted loggers as nature-destroying idiots. From that point forward it became impossible to have a reasonable conversation with someone who self-identified as an environmentalist. They now only sought our removal.

Their outrage was ill-conceived. The trees had blown over. They were dead. Wildlife could no longer pass through the area. This included the endangered native bull trout. The lodgepole pine had damned up Challenge Creek so tight in a couple places, the federal wildlife biologists who were temporarily staying at the Challenge Creek cabin feared the bull trout wouldn't be able to reach their spawning beds.

I learned about this one evening after walking over from my camping site to visit my neighbors. I was shocked to hear their proposed remedy. They'd settled on using dynamite to blow up the log jams. I quickly told them it would be easy to pick the trees out of the stream using a piece of logging equipment and do it with almost no silt entering the stream. They immediately became uncomfortable.

So I changed the bull trout subject into why these fish were in obvious decline. The reason was locally well known. The non-native brook trout were breeding with the native bull trout and creating a sterile hybrid.

I then presented my own theory. It seemed to me that one fish specie or the other was eventually going to go extinct. The surviving specie would be the one that could produce more viable offspring. They didn't answer. It was very apparent by their nervous demeanor that they feared revealing their own personal opinion.

I understood their trepidation. The van load of angry unreasonable people was a strong indicator that the environmental movement had ebbed into something sinister. The careers of these honest wildlife biologists were in jeopardy, because the federal bureaucracy was at that very time being infiltrated by political ideologues. Anyone not towing the party narrative (loggers bad and must be stopped) was being forced out.

Without pushback, the environmental/political ideologues continued stopping timber sales under the guise that the bull trout were declining solely because of silt coming off of logging jobs. Today, almost all the loggers are gone. Yet this native fish didn't bounce back and it appears that unless something changes quickly, the bull trout is destined for extinction.

……

In the fall of 1993 and in the dead of night, a terrorist associated with the environmental group 'Earth First' snuck onto our logging job and like all terrorists, he sought to advance a political aim through fear and intimidation. He proceeded to sabotage and destroy almost all of Dad's equipment, which included an IID (improvised incendiary device) that was ignited within the yarder. The skidder that I'd operated on the day of Dad's accident was a total loss. He also 'spiked' several logs (the band saw at the sawmill can explode when encountering the spike, dismembering

Dad's last trip to the forest (December 2014). The expressions on Adam (left) and Shawn's faces reveal the duality of the day, which fluctuated between joy and sorrow.

or killing a sawmill worker).

I showed up the next morning and discovered the damage and spray painted graffiti that the assailant (Ronald Constable, 'Earth First Ron') had left behind. His message was quite clear; he hated loggers.

Dad was working in a different area of this massive blowdown timber sale. I drove over and told him what had happened. I watched the light fade from his eyes. I contend it was at that precise moment he changed. It was mild, but he was never the same again.

……

Dad's company never recovered. He went bankrupt in 1999, joining hundreds of other logging families who'd been forced into a no win situation by the politically powerful/motivated.

The bank was more than eager when my brother, Shawn (U.S. Coast Guard veteran), and I showed interest in taking over some of Dad's payments. We soon acquired the few remaining pieces of worn down logging equipment and started our own logging company (I stayed in logging because of a vow I'd made after my own near death logging accident that had occurred on April 9, 1992, which was, "I'll find the solution or die trying.") The school of hard knocks therefore continued for us unabated.

Today, where there were once hundreds, my brother Shawn, nephew Adam and I are among the last of perhaps a dozen logging families still working the woods of northwest Montana. We survived by eventually inventing the world's first 'Tethered Yarderlogging System' (currently being studied by Montana's DNRC), which gave us an edge.

Ronald Constable was arrested a few months after his terrorist attack. I traveled to Helena in 1995 to testify on behalf of the federal prosecution Earth First Ron was found guilty by jury. The judge imposed a light sentence of time served and community service. We were then informed a civil suit was double jeopardy and Ron couldn't be sued. The last I heard, 'Earth First Ron' was working for the Oregon State Fish and Game.

Dad passed away in January of 2015 from brain cancer. The doctor said it was unrelated to the accident. A brain scan revealed a large void behind the impact site.

CHAPTER 3
FREE LAND

In the fall of 1985, my enlistment with the Marine Corps was over. I returned home to Montana and rejoined Dad's logging company. It had grown large and unrecognizable during my four years away.

My time in the Corps was spent with VMFA-112. This unique Fighter/Attack Squadron is home to the Marine Corps most experienced Fighter/Attack Pilots and RIO's (Radio Intercept Officer). It was a privilege and extremely motivating to be a small part of the maintenance element, which kept this Squadron of experts flying. Photo taken by my Marine Corps brother Joe Crowdus.

Preceding my enlistment, Dad had been a small gyppo logger. The summers and Saturdays spent logging with him while still in high school, were hard but happy. The experience had not only strengthened an already good father/son relationship, but cemented a friendship between us as well.

Dad's previous crew had been very familiar to me. It had consisted of Uncle Irving and two close family friends, Larry and Doug. All were hard workers and a pleasure to be around. Larry and Doug were natural comedians and spent every lunch competing for the next laugh. Uncle Irving was the intellectual of the group and always had something interesting to say. Unfortunately, those days were gone forever, as they'd all moved on to other endeavors.

Dad's new upsized company had every feature of good things to come. After all, it had grown perhaps five-fold and the prospects for future work seemed limitless.

Yet that first day back, I sensed something was off. I just couldn't quite put my finger on it. Perhaps it had something to do with all these stern and unfamiliar faces, some of which were giving me less than friendly sideways glances at lunch. The message transmitted by those glares were abundantly clear. I was to be treated as the perpetual outsider for committing the grievous crime of being the boss's kid. I therefore attributed that nagging pit in my stomach as simply missing the good old days, which I did. Yet, even with this self diagnosis administered and accepted, the feeling of dread persisted.

It took a few days before it finally dawned on me, 'that something' was Dad. It was so subtle, I doubt anyone who hadn't worked with him prior to the expansion could have recognized it. He wasn't happy. When I confronted him with this observation and suggested that perhaps getting big wasn't all it was cracked up to be, he laughed it off as ridiculous. So I smiled and played it off as a joke… but it hadn't been funny.

I knew there was no going back. Dad had leveraged everything he and Mom owned for this chance at something 'better'. So it was all or nothing and a naysaying son wasn't helping matters. So I climbed onboard the departing train, bound for destinations unknown and adopted the outward appearance of promise and hope, while struggling inwardly with a festering sense of murky foreboding.

……

Dad hadn't been the only small gyppo logger that had recently gotten the opportunity to grow. The reason had everything to do with efficiency or lack thereof. A large industrial timber company, Plum Creek, had recently acquired the railroad land holdings of Burlington Northern. (The railroad ownership has transferred several times over the decades, from Great Northern to Burlington Northern, to Plum Creek to Weyerhaeuser and most recently to Southern Pine Plantations). Plum Creek deemed the last remaining railroad company crew (Royal Logging) inefficient and subsequently disbanded it. To fill the void, several small family logging operations, including Dad's, were given the opportunity to expand.

At the time, we gyppos liked to imagine that we'd outcompeted the Big Company logging crews and that's why they went out of business. The truth was far simpler. The easy wood was gone.

To better understand the circumstances that allowed Dad and the other small gyppo families a chance at upper mobility, we must go back several decades and revisit how industrialized logging first came to the Flathead Valley, along with the striking differences between company and gyppo loggers.

The ecological ramifications of this region's unique geological placement must also be considered. Northwest Montana is on the wrong side of two rain shadows. The Cascade Mountain range in western Washington and the Bitterroots, which form the western border of Montana, conspire to minimize the amount of precipitation that can reach this isolated mountainous region. This lack of water, coupled with a short growing season, reduce tree growth to a painfully slow process.

When the railroad first entered the area in the early 1890s, it found the valley floors of the Flathead Valley and the adjoining Smith Valley almost treeless. The surrounding hills and narrower north-south valleys however, were teeming with large old growth timber.

'Large' tree is a relative term in this case. For example: The area sandwiched between the Cascades and the Pacific Ocean is an anomaly in the west, because of its high annual rainfall. This ample water and long growing season allowed the old growth of this coastal region to attain staggering heights and girths. I'm unaware of the overall combined average size of those old trees, but a diameter of 8' seems a reasonable guess. Conversely, here in northwest Montana the trees were equally ancient, but the size of the old growth trees aren't even close to being comparably proportioned, perhaps averaging less than 30" in diameter. Again I'm guessing, but you get the point; trees are much smaller here.

In this case, smaller was a good thing, because the Montana old growth was still somewhat of a manageable size (coastal old growth was often dynamited in order to create smaller movable pieces). Another positive for the early railroads was the abundance of Western Larch in Northwest Montana, which made premiere railroad ties. The railroads were desperate for those ties because of the ongoing massive westward expansion.

The Great Northern Railroad soon began punching spurs anywhere feasibly possible, while bypassing the difficult areas and slopes. To use logging slang, 'they were juicing', taking the easy timber, on the easy ground.

The railroad companies employed large 'company' logging crews, tasked with harvesting along the railroad right of ways. They used two man crosscuts for felling and segmenting the tree into manageable lengths (logs). Horses pulled the log to the tracks, where they were loaded onto flat cars. The logs were considered 'delivered', when they splashed into the northern waters of Flathead lake (today, those old railroad pilings are supporting the North Valley Yacht Club docks).

The railroad town of Somers, located on the north shore of this large lake, existed for one purpose. Fish those logs out of the water, mill the logs into railroad ties and treat the freshly cut beams with creosote. My great-great-grandfather Averill worked at the Somers sawmill and yes, today it's a superfund waste site.

The railroad company crews couldn't keep up with demand, which created opportunity for the gyppo. There were almost no roads into the mountain valleys at this time, but several fairly large rivers flowed into Flathead lake. The logs could therefore be floated to the Somers mill.

The bigger rivers such as the Swan, were utilized early on, but as the cut continued upstream, the water became shallow and unable to float the logs. The gyppo however, out of necessity, must be inventive or go broke.

Henry Good was just such a gyppo. My step-great-grandfather Faye Moses worked for Henry during at least one season in the 1920s. Decades later, sometime late in the 1960s he told me the story of how they 'delivered' their logs down the Stillwater River to the lake.

I sat there transfixed, not only with his story, but by what he was doing with his hands. Using a pocket knife, he was peeling an apple, in a manner I've never seen replicated. Starting at the top, he cut a thin continuous strip of peel as he slowly rotated the apple in his hand. The ribbon of equal width and free of any white fruit, was still short as his story began.

He told me, "In late summer, they traveled on horse back, 30 or so miles north of Kalispell along the Stillwater River until they reached that seasons' harvest site. The first task after setting up camp, was constructing a dam across the creek (later named Good Creek). The low water of late summer made the operation fairly easy. Once completed, the dam slowly backed up the water, creating a reservoir.

The next phase was cutting timber. Each log was clearly stamped on both ends, with Henry Good's unique identifying brand and the year it was harvested. The logging continued until winter had frozen the man-made lake solid. Using horses, they skidded the logs onto the ice. When the last log was placed on the frozen surface, they went home and waited for spring.

They returned to the logging site just before the spring runoff reached its zenith. When Henry judged the critical moment had arrived, he used dynamite to blow the dam. The logs and associated debris quickly flushed downstream with the initial flow, but logs soon began getting stuck along the banks. They hurriedly chased and herded the logs downstream, freeing the many beached logs and breaking up the inevitable log jam. The clock was ticking. The water levels would soon begin dropping as the spring runoff slowed.

The journey lasted a few weeks and ended when they finally reached Flathead Lake, but try as they had, many logs never reached this final destination. Every log that didn't make the trip, which included many 'sinkers' (many remain visibly submerged under the Stillwater), reduced the future paycheck.

The lake logs were eventually gathered into rafts, all mixed in with logs from other gyppo loggers working different drainages. As the logs patiently floated awaiting to be sawmilled, more sank. When the logs were finally brought out of the water, the stamp was recorded along with the year. The log was scaled and a paycheck sent to Henry Good. He dispersed the gyppo wage to his crew, after deducting any outstanding 'draws'.

The story had come to an end, just as he finished peeling the apple. He caught me staring at the long curly red ribbon of apple skin. He winked at me and slowly began lowering the unbroken and continuous peel onto the table next to him. I marveled at its reassembled appearance. Minus the stem, it appeared an intact apple.

......

It can be readily assumed that Henry Good lost a large portion of his crew at the end of each season. Many undoubted took their hard earned logging experience and sought employment with the well funded railroad company crews, who by comparison were living utopian lifestyles with regular paychecks.

Undeterred by hardship, the ever-persevering gyppo Henry headed back up river each consecutive year as summer began to wane, with a 'new' crew in tow, many undoubtedly fresh off the train and desperate for employment. They stopped just a little further up stream from the previous year's cut and began building a dam.

As for my step-great-grandfather, he and my great-grandmother Ruth had saved enough money for a down payment on a taxi business. It was short lived. After the stock market crash of 1929, few had money for a cab ride. They were unable the make the payments and the bank repossessed the taxis.

The Wall Street collapse and subsequent economic Great Depression also ended the era of the Northwest Montana river gyppo. Henry Good retired from logging and began focusing on his ranch, located ten miles north of Kalispell, passing away in 1944.

......

Mom's family had come to the Flathead Valley early. Dad's side was on its way, but was taking a different path through Montana.

In 1910, the federal government doubled the homestead allotment from 160 to 320 acres. My great-great-grandmother on the Henry side had several married daughters and one unattached son, Ora. They all decided to leave Kansas and go get some of that free land in Montana. They would learn the hard way that nothing is free. They settled in eastern Montana near Glasgow, in an area soon known as Kansas Flats.

Granddad Bruch was less then a year old when the family moved onto this treeless arid prairie. Much of what I know about this era was passed down from Grandad, through Dad and eventually to me. I do however have another invaluable source. A book written by Emma Hedman, 'Plow a New Furrow' chronicles her family's immigration to the United States from Sweden in 1893.

They came for the explicit purpose of getting some free land. Her father, mother and oldest sister Anna, who at the time was their only child, reached their destination on the Great Plains later that year in Nebraska, and claimed a 160-acre homestead. (Anna would eventually marry Ora.)

This first dryland farm ended in failure. Her story tells of several more attempts at various locations throughout the northern Great Plains states. Each time, the growing family gave every ounce of effort with unwavering and undiminished hope, believing eventually they'd succeed at dryland farming. The one in North Dakota came oh, so close.

In the autumn of 1914, all the area farmers were working together, harvesting a bumper crop of wheat with a co-op thrasher moving from one farm to another as quickly as possible. The Hedman farm was next when a giant hail storm swept through the area, knocking down and destroying the wheat. The children, unaware of the dire consequences, gathered up some of the hail and asked their mother to help them make ice cream.

The Father (Peter) had enough of dryland farming. He sold the homestead for $25.00 an acre and moved his family to Wisconsin, which is east of the 100th meridian (imaginary line running north from the Texas panhandle up through the middle of North Dakota) and its wetter and more hospitable climate.

In 1916, the Hedmans decided to give it another try, but not as dryland farmers. This time they'd try to raise cattle in Montana, and as fate would have it, they homesteaded on Kansas Flats. Our two families eventually combined when Emma's older sister Anna married Uncle Ora. The scope of Emma's story expands to include her new extended family.

The next ten years tells a story of endless struggle and hardship, with every member of the family doing whatever they could to get through the next obstacle. The shared tribulations formed unusually strong family bonds.

1920 Wedding photo of Uncle Ora Henry and Aunt Anna Hedman-Henry. Photo provided by Carrie Henry-Ramsey (Great granddaughter of Ora and Anna)

Although difficult, the many families living on Kansas Flats felt they were slowly succeeding. The Bruches were raising horses to sell. The oldest son Kenneth (Granddad) had a natural talent for bareback bronc riding. He dreamed of competing on the rodeo circuit.

Then, years of drought descended in the 1920s. It was the beginning of end for the Kansas Flats homesteads. The dryland homesteaders just didn't know it yet. They'd grown so acclimated to hardship, they simply assumed it was just another obstacle in a never-ending line of challenges.

What you won't read in Emma's account is any fault directed towards the federal government for her family's persistent hardship and failure. Maybe she would have added something had she known about John Wesley Powell's attempt to give them a fighting chance.

In 1890, before the throngs of would-be homesteaders began flooding into the Great Plains, Powell proposed drastically increasing the individual allotment to 2500 acres. Powell had been sent west by the federal government following the Civil War. He led several scientific expeditions for the explicit purpose of achieving a better understanding of the arid landscape, with the ultimate goal of discovering the feasibility of introducing European style agriculture.

Two decades of intense study in the west had transformed him into the leading scientific expert on the subject. Modern science and scores of scientific fields sprout from these early western expeditions.

In 1890, the federal government was making the final plans that sent swarms of dryland farmers into the parched western landscape, each with the promise of 160 acres. Powell quickly recognized the flaw in the plan. The 160-acre allotment was horribly insufficient and would only lead to unnecessary hardships for the arriving hopefuls and potentially give birth to a whole host of other catastrophic consequences.

Motivated by a desire to avert an impending disaster, Powell set up a meeting in Washington D.C. with a couple of high-ranking political appointees. He began by telling them that 160 acres wasn't enough for a family to succeed economically. When asked how much was needed, he immediately responded with 2500 acres. The politicians balked at such a high number, pointing out that the homestead act of 1862 was nearly 30 years old and 160 acres had always been ample for eastern farmers.

Powell tried to impress upon them the desperate lack of water west of the 100th meridian, but to no avail. A heated back and forth ensued with neither side backing down. Frustrated with the bureaucrats, Powell suddenly changed the subject. He blurted out, "If the practice of frequently burning the landscape in a manner consistent with Native American methods isn't continued, the western forests will one day explode into uncontrollable conflagrations." This outburst not only ended the meeting, but John Wesley Powell's formerly unblemished career. The politicians soon branded him crazy and removed all future government funding.

Today, the average farm size in eastern Montana is 2500 acres and the mega fires throughout the west grow worse with each passing year. I make this point to illustrate, that had the politicians listened to Powell and sought the scientific path instead of the political, a deeper understanding of the western forests may very well have been found and our current ecological crisis may not have reached the tipping point.

That however, is not what happened. Instead, the incident marked the beginning of a disturbing political trend. Politicians with preconceived and overly simplified political narratives began pushing their own biased views upon nature and the populous, under the guise of science. If an intellectually honest scientist such as Powell disagreed with the narrative, they were quickly removed and replaced with a scientific 'yes man' whose loyalty could easily be bought with a lucrative position of prominence.

Sadly, Powell was far from the last scientist destroyed by a political system, which has not only proven itself incapable of solving real world problems, but is the very source behind every festering ecological issue we now face. In this current era of dire and looming ecological crisis, it's time to state the obvious. Winning a popularity contest doesn't suddenly endow the political victor with an intimate knowledge of everything. The dust bowl era is a good example of this type of political incompetence and arrogance. It remains one of worse ecological disasters of all time and could have been avoided if the politicians would have listened to Powell. He was a person, who unlike the politicians, actually possessed practical experience within the affected region.

Yet even though the occupation of 'professional politician' is responsible for causing this ecological calamity, it's never held accountable. The powerful bureaucracy was therefore able to continue down its destructive path, making countless other poor decisions. The negative ecological consequences have snowballed and are now threatening to destroy life on Earth.

'We the People' must quickly install a new way of communicating with our elected officials. It needs to be one that reduces the bureaucracy to its proper clerical positions, while simultaneously harnessing the intellectual might and enormous empathy of the American people. The technology for this remedy already exists.

This envisioned internet-connected advisory group should be composed of citizens and scientists possessing high levels of practical experience on specific subjects. They must be free of direct monetary conflicts of interests.

This 4th branch will provide invaluable insight for any upcoming decisions being made by our elected officials, with all interactions occurring in full view of the American people. The creation of this new age social advisory platform will finally make it possible to eliminate the corrosive effects of money in politics and allow the 'currency of thought' to become the true influencer.

After its installation, special interest groups can make their case directly to this American forum and receive an answer from 'We the People'. This will end today's practice of the well-connected and wealthy offering support and/or campaign contributions for favorable consideration, which is without a doubt the driving force behind all of today's legislative malpractice.

Unfortunately, the technology for this potential solution didn't exist for my early Montana ancestors, who were offered only two choices by the federal bureaucracy: survive on a harsh dry landscape or die trying.

……

By the mid 1920s, a drought had caused the grass to stop growing. As a result, the Bruches needed every bit of land they owned to graze the horses during summer. This tactic still proved inadequate. The horses were losing weight. The problem was only going to get worse, because there wasn't going to be any hay, which meant no winter feed. The horses were going to die of starvation. A solution had to be found.

I'm unaware of many of the details as to how the Bruches came to know a bachelor rancher in Bigfork Montana in the Flathead Valley, other than he was terminally ill and had offered a proposition. If the Bruches took care of him until he died, he'd leave them his 160-acre ranch. Again, I'm unsure of all the aspects that followed, but it seems likely that great-grandma and great-grandpa Bruch went over on the train with their younger children (she had 7) to nurse the ailing rancher in the fall of 1926. The old cowboy passed away three and a half months later.

There wasn't any money to ship the horses by rail. So Grandad Bruch, who was 16 years old at the time and his next younger brother 'Judge' (14 years old, I think) rounded up the horses and with the chill of fall in the air, began herding them towards the Bigfork ranch and the awaiting hay.

The several hundred mile journey across the relatively flat plains was purposefully slow. The malnourished horses couldn't be pushed and needed extra time to graze. Eventually, they reached the Continental Divide and descending Marias Pass with the high mountains hemming them in on both sides.

Luckily, it was still three years before a massive forest fire swept up this highly incised valley (mentioned in chapter 2), which would have likely made it impossible to find enough grazing grass as they traveled the scores of miles down the Middle Fork of the Flathead River that constitutes the southern border of Glacier National Park. As it was, grass was likely plentiful.

They eventually reached the Flathead Valley and continued following the river as it abruptly turned south. They herded the horses a further 20 miles on the east bank of the river, until Flathead Lake was just a half mile away. They turned left and rode up and over a hill. A quarter mile more finally put them on the ranch.

Grandad and his brother delivered the horses, but this was only the halfway point for them. They immediately had to return to the homestead or risk having it deemed abandoned by the federal government, which would cause the family to forfeit ownership.

They walked twenty-five miles north, to Whitefish and the railroad that intersects this small town. There wasn't money for a ticket, so they jumped an eastbound freight train. When the train reached the plains, a sudden cold snap hit, plunging temperatures well below zero. Granddad would later tell Dad that he'd never been colder in his life and thought they were both going to freeze to death. After the train ride, they walked to the homestead and survived the winter, satisfying government occupancy requirements.

They had hoped the drought would end the following year and allow them to bring the horses back, but it wasn't to be. The lack of precipitation persisted year after year. A few years later, the Great Depression began, which reduced the value of the horses to almost nothing. The 160 acres in Bigfork couldn't support all the horses, so many were sold at near give away prices.

They still weren't quite ready to give up on the homestead just yet. Members of the Bruch family rotated with one another, traveling back and forth to maintain the occupancy requirement. Most of the other families had abandoned Kansas Flats back in 1926, never to return. They had also come west and relocated throughout the Flathead Valley. The last attempt to save the homestead was made in 1931. It was during this last effort that Granddad went on his first date with my future grandmother, Barbara Stalcup.

I can't tell you what they talked about or where they went, but I can tell you how the date ended. The car had overturned and was sliding down a gravel road. In this era before seatbelts, Granddad put his right hand down on the still moving road surface in a desperate attempt to keep from falling out. When the car finally came to a stop, the skin on his palm was gone.

He was rushed to a small town hospital for surgery, but in this era before skin grafting, there was only one option. He woke to the startling discovery that his right hand had been amputated. And yes, he was right handed.

This early 1930s photo of Granddad, shows him seated in a place where he always felt at home. If you look close enough, you can see that this photo was taken after the accident.

After healing, he attempted to ride the bucking broncs, but found he no longer had the necessary balance to compete at the higher level, dashing his rodeo dreams. To make matters worse and because it was the depths of the Great Depression, prospective employers wouldn't give a one-handed man a second look. The only job he could find paid next to nothing, but without options, he took it.

The next two years saw him living an isolated existence on the Great Plains, watching over a flock of grazing sheep. When it came time to move the wooly animals, he did so from the seat of a Conestoga style wagon, which doubled as his sleeping quarters.

He fell into a deep depression and at some point told his mom how he felt. She responded in a letter that Dad found in the 60s after Grandma Barbara's untimely death. The main theme of the letter was a simple message, designed to get her son to wake up to his new reality. She said, "You're half a man now and should expect half a wage." The letter so enraged my father, that although great-grandma Bruch lived on into the 1980s and was only 20 miles from us, I never met her. When Dad asked Grandad how he felt about the letter, he merely shrugged and said, "It was a different time" and never showed the slightest bit of anger towards his mom.

In Great-Grandma's defense, her seemingly cold hearted message didn't destroy her son. He went on to marry Grandma Bruch in 1934. That was followed a year later by the arrival of their first child Aunt Virginia, with Dad showing up four years later in 1939. Those who knew Grandad, universally described him as a remarkably strong, capable and caring person. If the word 'crippled' entered the conversation, it was merely to marvel at why this descriptive word didn't seem to apply to him.

Roosevelt's New Deal helped many struggling families during the depression and like so many others, my grandparents were eternally grateful to FDR. They would remain loyal to his memory and the Democratic Party for the rest of their lives, blissfully unaware that although it hadn't been FDR personally, it was others employed within his occupation (politicians) that caused them all so much heartache and pain.

Granddad was soon working for the WPA (Works Progress Administration). He was immensely thankful for the job, but soon became

1934 wedding photo of Grandma (Barbara Stalcup-Bruch) and Granddad (Kenneth Bruch).

witness to and later lamented the horrible inefficiencies of government. One of Granddad's jobs was mowing grass with a horse drawn mower. He later described the less than efficient use of manpower as, "One coming, one going, one shitting and one mowing."

……

It took World War II to finally put an end to the Great Depression. All of a sudden and because most able-bodied men were away in the military, it was easy to find a job (Grandad was exempt from the draft because of his missing hand).

Lumber was a war material in high demand. Uncle Ora Henry was by now a well-established timber faller, having landed the rare logging job during the depression. Ora and Granddad teamed up to saw timber, each pulling an end of a two-man crosscut.

I never met Ora, but I'm familiar with stories about him. He was a WWI combat veteran and a man of few words. He also possessed a strong reputation as someone you could count on in desperate times. He wasn't known as a jokester, but one story I've heard repeatedly was of a rare exception to his otherwise stoic composure.

On an extremely cold morning as Ora and Grandad were walking to their 'strip', Ora abruptly stopped and looked down with a worried look on his face. Grandad surprised by the sudden movement, quickly asked him what was wrong. Upon which, Ora looked up and over at Granddad with a look of immense relief and said, "For a minute there, I thought I forgot to put my pants on."

It's far from hilarious, but gives us an insight into a time when moments of levity were often few and far between. This rare occasion became a memorable event, deemed worthy of passing down.

……

The war eventually ended and service members came home, each with a desire to put the war behind them and get on with their lives. A housing boom soon exploded, creating a huge demand for construction lumber.

Following the abandonment of Kansas Flats, Uncle Ora, Aunt Anna and their three children (Melvin, Elwood and Vernon) had moved to Kila, located 10 miles west of Kalispell. The Bruches had gone to the Bigfork Ranch, which is 15 miles south of Kalispell. In order to be closer to his sawing partner, Granddad and Grandma, Aunt Virginia and Dad (Marlowe) moved to Kila in the mid 1940s. Uncle Ora and Granddad however, would soon stop cutting the large old growth and begin focusing their attention on the small trees found within the second-growth forest.

……

They were part of a little known gyppo movement, started by an unpedigreed people. This next generation of gyppo innovators didn't possess the cutting edge logging technology being utilized by the big company crews. This sizable obstacle however, was more then offset by being free from any type of government/political interference.

This paradigm shift consisted of a scattering of a few families working independently from one another, but sharing a common vision. They were going to attempt to seize upon an opportunity, in which the big companies and government saw none. Most were starting from near zero, but would go on to change everything.

……

The forest that regrew in the decades following the removal of the old growth was very different from the ancient forests it replaced. The previous old growth forest had been 'park like' with an open grassy forest floor and dominated by 'long living' fire-resistant tree species (Western Larch and/or Ponderosa Pine), with their lower limbs perched high above the ground.

Within this second-growth forest however, not only were these majestic trees species no longer a dominating force, but frequently absent. The Inland Douglas Fir and other shade-tolerant tree species had taken over command of the landscape and were now in firm possession of these fire-prone sites. This new forest had grown unnaturally thick.

To understand why the transition occurred, we must go back in time to the 2.6 million-year-long duration of the Pleistocene ice age.

THE FOLLOWING IS A BRIEF SYNOPSIS, WITH A BROADER EXPLANATION IN CHAPTER 5. THIS IS AN ORIGINAL THEORY...

When imagining an ice age, vivid images of intense cold and glaciers comes to mind, which is accurate, but many aspects of that long ago landscape weren't necessarily wet or frozen; in fact some places were quite fire prone. This reality is just one of the many truths that can be gleaned by looking deep within the individually-evolved characteristics found within the many tree species of the northwest.

For instance, the cold/snow adapted Alpine Fir grows in high elevations and is shaped like a slender pyramid, which helps it shed heavy snow loads. This thin barked, shallow-rooted tree is absolutely fire intolerant and if ignited, explodes like a roman candle. It also doesn't produce pitch to heal wounds, is shade-tolerant and has a fairly short lifespan. Incredibly, this tree along with most other shade-tolerant trees, doesn't require exposed topsoil for regeneration.

When all these clues are considered, a fairly clear picture begins to emerge into how this particular tree specie existed during the Pleistocene. Obviously, it grew in areas where fire wasn't a possibility, which is evidenced by the fact that this tree never developed any fire-resistant traits.

There are several other comparable tree species found within the northwest with a shared commonality of shallow roots, thin bark and shade tolerance, which likewise lack the ability to survive even a low-intensity ground fire. Although similar, these differing tree species clearly spent extended periods of isolation separated from one another with each developing the uniquely evolved characteristics that allowed them to thrive in their own particular circumstance (elevation, excessive snow load, saturated soils, etc.).

Going forward I will refer to this type of tree as 'Glacial', because they resided and survived in very wet and/or cold environments, such as near the ice flow.

The other type of tree shall be referred to as 'Inter-glacial', which is in almost every way the polar opposite of the 'Glacial' tree. They possess thick, scaly, heat-shielding bark, deep tap roots, extended life spans and produce pitch to heal fire wounds and other injuries. They require direct sunlight, which causes the lower limbs to die and fall off when cast in the shadows. Unlike the Glacial tree, they need exposed soil for rejuvenation, such as that provided by a low-intensity ground fire.

The evolutionary story of how these magnificent trees evolved (Ponderosa Pine, Western Larch, Sequoia and others), can't be told without including several members of the now extinct mega fauna (mastodon, mammoth and others).

I contend that these floral and faunal species were components of a robust ecological mechanism, which was so powerful, it was capable of influencing the global climate. This influence can clearly be seen in the Vostok Ice Core Analysis graphs.

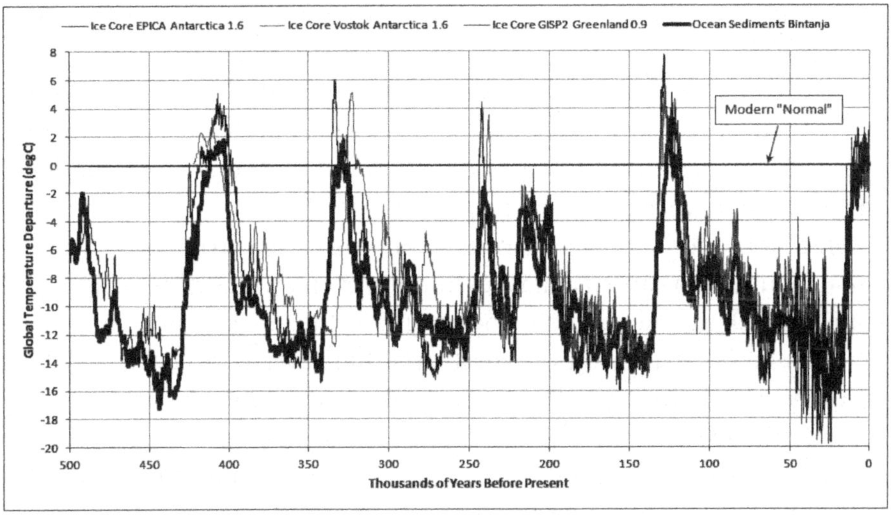

I contend the rapid up and downs represented on the graph were the product a robust ecological mechanism, which was once powered by countless floral and faunal species.

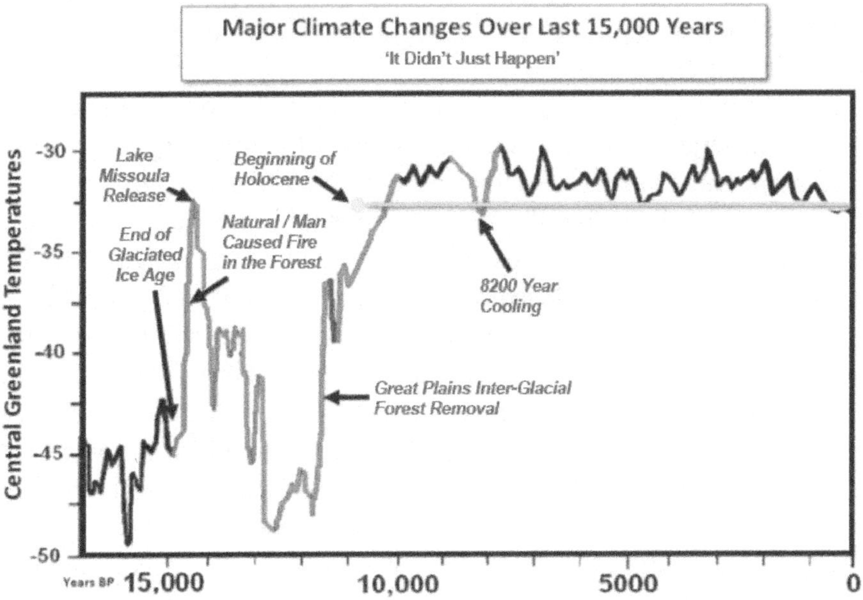

I also contend that at the 14,800 mark, Man began misusing fire in western North America, which ultimately lead to a massive flood event (Glacial Lake Missoula). The 11,600 year mark is where Man removed the Great Plains Inter-glacial forest and caused a mass extinction event. In the absence of the mega fauna, the global climate collapsed, which is depicted by the fluctuating flat line that followed (Holocene).

Focusing on the 14,800 year point on the graph, something happened that proved to be the beginning of the end for the Pleistocene. I believe it was at this moment that man began destroying that climate mechanism with the misuse of fire. These incidents occurred in the western portions of North America and subsequently led to the exaggerated development and eventual release of Glacial Lake Missoula.

Three thousand years later, using the same fire tactics on the once-forested Great Plains, man caused a mass extinction event, which ultimately wiped out almost all the mega fauna. The bison was the sole survivor.

This mass extinction event destroyed the climate mechanism, which abruptly ended the 2.6 million-year reign of the Pleistocene ice ages. The several 100,000-year ice age cycles that are seen on the graph prior to the collapse gave way to a flat line. The following period is known as the Holocene, which witnessed the formerly massive glaciers slowly recede and eventually expose today's Canada. The following era saw the dramatic expansion of deserts in the southwest (Altithermal).

Several millennia after the collapse, starting about 5000 years ago, the Native Americans began successfully developing beneficial burning strategies, which gave preference to grass. These methods permitted the bison numbers to slowly rebound on the now treeless Great Plains. The modern bison are very different from their Pleistocene predecessor, having reemerged as a much smaller animal with curved horns.

It's impossible to know exactly where the Native Americans first began developing and deploying the beneficial burning methods, which led to the restoration of the grasslands. Evidence of this ancient activity however, can still clearly be seen in Northwest Montana. As previously stated, when the railroads first entered the Flathead, this large valley and the adjoining smaller east-west Smith Valley were mostly treeless, grass-covered prairies.

This ancient deforestation was made possible because the prevailing winds could be utilized to push the fire across the forest canopy. Once removed, repeated maintenance fires over the course of several millennia removed any evidence a forest ever existed on these valley floors, and also kept trees from infiltrating these now grassy plains. The grass maintained a population of bison and other wildlife, which provided a sturdy food source for the Native Americans.

The north-south valleys and adjoining hills however, remained timbered. This occurred because the slopes deflected the strong prevailing winds and inhibited the forest-removing canopy fires from being used to full effect. This proved to be the positive turning point for the arid forests, which are located here and there throughout the west. The Native Americans had accidently stumbled onto a method of recreating an anemic version of the long lost Pleistocene's fire resistant-forest structure.

Where previously the mastodon had played the indispensable role in maintaining a separation between the two incompatible tree types (Glacial and Inter-glacial), the anthropogenic burning would prove to be an adequate substitute for accomplishing the same division by creating a trial by fire, which subsequently forced the Glacial trees to reside within areas of elevated moisture. The frequent burning also permitted the fire-resistant Inter-glacial trees to dominate the fire prone aspects of the landscape. Over the course of the next several millennia, the Native Americans became experts at shaping their ecological surroundings with the wise and timely use of fire, and became the critical stabilizers of the global climate.

By 1890, John Wesley Powell recognized the importance of the Native American method of frequently burning the landscape. Unfortunately, he wasn't aware of the deeper story. He therefore couldn't adequately explain why it must continue.

The Forest Service was established in 1905, with Gifford Pinchot as its first chief. When asked about the Native American method of frequently burning the landscape, he derisively referred to the practice as 'Piute' forestry. He firmly believed the Native Americans were relicts of the past. Pinchot was horribly mistaken, but as a powerful political appointee endowed with the power of decision making, it was his opinion that mattered. Influenced by his European education, Pinchot forced his own ideals on a landscape he personally knew very little about, and arrogantly adopted a nationwide fire-suppression policy.

The following decades saw Glacial trees quietly begin infiltrating the old growth forest. As the shade-tolerant trees grew, they began creating fire ladders into the high and flammable branches of the ancient sentinels of time. This shading of the forest floor thinned or completely removed the grass.

Something beyond the obvious was happening within the Native American-created fire-resistant forest structure. It was very subtle and has everything to do with the evolved functionality of the two opposing forest types. A Glacial forest removes soil moisture, whereas an Inter-glacial forest protects and builds a reservoir of water within the forest floor soils. This water reserve does far more than elevate fire protection for this forest type. It also influences weather patterns in a very positive way.

In Chapter 5, I will expand on the Pleistocene climatic mechanism, Lake Missoula flood, mass extinction event and our current climate oscillations, while hinting at a revolutionary new industry I've devised. It is an industry designed to restore climate-stabilizing, fire-resistant forest structure.

CHAPTER 4
TRIAL AND ERROR

Fire suppression throughout the western landscape was established long before the federal government officially adopted it in 1905. It had begun decades earlier in the wake of the Civil War, as settlers began flooding west. These strangers to an unfamiliar land initially saw fire as only destructive. So with the best of intentions, they encouraged the Native Americans to stop burning the landscape, sometimes by force.

The subsequent unnatural reactions that took place within the western forests along with massive swaths of global deforestation is the true driver of climate change.

RECAPPING AND BUILDING ON AN ORIGINAL THEORY

The 2.6 million-year-long Pleistocene ice ages produced radically different tree types. For simplification, I've placed these divergent trees into two separate groups, based on shared similarities. The unifying factor being whether the tree specie 'possesses' or 'lacks' fire-resistant traits. I've named the former 'Inter-glacial' and the latter 'Glacial'.

During the Pleistocene, the Inter-glacial trees and mega fauna co-evolved to become a powerful climate-influencing machine. Starting about 14,800 years ago, man began destroying that powerful mechanism with the misuse of fire. By 11,600 years ago, this continued activity had caused almost all of North America's mega fauna, excluding the bison, to go extinct (mastodon, wooly rhino and many others).

This shattering of the Pleistocene mechanism caused the global climate to unravel. Over the next several thousand years, the massive glaciers covering what today is Canada slowly melted.

The constant meltdown produced an enormous body of water known as Lake Agassiz, formed within the waning ice fields of northcentral North America. This giant lake is believed to have filled and emptied a total of four times. The first lake phase began 12,875 years ago and is known as the Lockhart phase. It lasted about 300 years before empting in a violent outburst.

The next lake immediately began filling (Moorhead phase) only to suffer the same fate about 900 years later. The third took a 1000 years and the last one about 1500 years. This last Lake Agassiz is named Ojibway and disappeared 8480 years ago. At this time, Canada was ice free for the most part.

The period that followed is known as the 'Altithermal' and existed between 8000-5000 years ago. It was a horribly dry era that witnessed the North American southwest transitioning into deserts. Bison teeth found within the fossil record from this timeframe show extreme wear, suggesting these animals were consuming dry, grit-covered grass.

The dryness of this epoch produced massive dust storms, which caused shifting dirt dunes to become a normal feature on the Great Plains. Water was scarce. Populations plummeted. It was a terrible time to be alive.

Then 5000 years ago, the rains began returning. The bison numbers slowly rebounded. Today, our current scientific understanding still cannot explain why the climate began stabilizing.

I contend it was at this time that Native Americans began developing beneficial burning strategies, which ultimately led to the restoration of the Great Plains grasslands. This reaction east of the Continental Divide was likely a direct response to the initial restorative burning taking place on the west side, which successfully began reinstating the separation between the two incompatibly-evolved tree types.

The restored division by fire enabled the many differing tree species to begin growing on aspects of the landscape that were more attuned to their own specifically-evolved characteristics. Each forest type soon attracted other floral components from the 'long lost' Pleistocene mechanism. The two distinct plant communities reformed and stabilized. Ecosystems rebounded and started flourishing.

This man-influenced forest structure was only a weak representation of the previous Pleistocene mechanism. However, it proved strong enough to influence the climate in a very positive way and explains why the rains began returning to the Great Plains and forced the Altithermal into dormancy.

......

Global populations began growing. Driven by agriculture, this expansion soon saw settlements springing up worldwide. Agriculture initially began popping up in the aftermath of the mass extinction event, indicating food source desperation. The steady population growth that took place after the Altithermal went dormant, eventually required management, which unintentionally gave rise to strong centralized authorities.

Some leaders were benevolent, many more were not. These powerful few, with various amounts of influence, frequently passed down this authority to their heirs. This resulted in the rise of the pharaohs, with later dynasties cropping up around the world. Few, if any, of these families credited agriculture for their lofty societal positions. Instead, they often explained away their hold on elevated power as being divinely obtained. Therefore, they saw it as their God-given right to control the fate of those under their rule. War and/or famine always followed.

America was to have been different. The U.S. Constitution was created to end birth-right rule, while also preventing a few people from centralizing authority onto themselves. The tragic fact, that slavery was permitted to continue after ratification provides evidence that something is missing from the Constitution. Furthermore, this well-intended document lacked guidance on how to manage the Industrial Revolution, which sprang up just a few years after its adoption.

Unfortunately, this void was filled by the political party system, which is guilty of overthrowing the original intent of America by allowing politicians to become the new ruling class. Complex problems brought on by the Industrial Revolution created a situation that allowed these new era pharaohs to make decisions well beyond their own realm of understanding, leading to a wide array of negative consequences. The ever-increasing industrial footprint caused a plethora of poor political decisions to become magnified.

This poor industrial age decision making is exemplified in the federal government's 1905 adoption of fire suppression, which officially ended the Native American climate-stabilizing tradition. Almost all 'civilized' nations on Earth followed our example.

More on the climate later. Let's dig deeper into how the bureaucracy got us into this mess. But as we go forward, keep in mind that water vapor is a far more effective green house gas when compared to CO_2.

......

In the aftermath of Gifford Pinchot's 1905 decision, billions of Glacial trees in the western US, which don't require exposed topsoil for rejuvenation, began infiltrating the moist floors of old growth forests. The Glacial tree had evolved within very wet areas of the Pleistocene landscape. As a result, they developed into very effective water pumps. This is polar opposite to the drought-resistant Ponderosa pine that can actually stop transpiration on hot summer days, which enables this Inter-glacial tree specie to conserve soil moisture when the forest needs it the most.

In the wake of fire suppression, the young Glacial trees growing under the old growth began withdrawing more and more water out of the upper layers of soil and expelling it through their needles into the atmosphere. This elevated fire risk by lowering forest floor humidity. These efficient pumps were also intercepting the water before it could reach the much deeper tap roots of old growth fire-resistant tree species, causing drought stress.

The lack of frequent burning also marooned the Inter-glacial seed cones on top of the duff, which unlike their Glacial counterpart, require exposed topsoil or some sort of aggressive disturbance for successful germination to take place. The problems for the western forests had only begun.

......

Another decision made by Pinchot in 1905 caused the Forest Service to perpetually ignore the transition and the growing threat it represented. He adopted European style tree farming as a one size fits all approach for all of America's forestry needs. This method by design, ignores and removes whatever forest structure had previously resided on the landscape and replaces it with a monoculture man-made version.

This 'scientific' forestry method was devised by a German named Johann Hienrich Cotta. He outlined this industrialized doctrine within a book he wrote in 1817, titled 'Advice on Silvaculture'. The premise of this method is simple and straight forward. Remove the existing forest and replant the land with a native tree specie, selecting one or more possessing suitable characteristics for shipbuilding.

The seedlings are planted in a grid like pattern. This enables each tree to receive adequate sunlight and water. Periodic thinning of this man-made forest maximizes growth rates by minimizing competition. When the trees reach marketable dimensions, they are all cut down and the process repeated by clearcutting.

At the time Cotta wrote his book, Europe's forests lay in ruin, caused by several centuries of aggressive ship building. The second growth that had grown up after the removal of the ancient forests were stunted and overcrowded. Within these thickets of small and very old trees were few stems large enough for ship building.

To clarify, Cotta's vision didn't seek to understand the mysterious underpinnings of that long gone ancient forest, nor did he ponder on how to restore what was lost. His entire focus was regrowing marketable sized trees as quickly as possible.

Gifford Pinchot was the first American to receive a forestry degree from a European university. When he achieved this distinction in the late 19th century, European style tree farming was considered the height of scientific advancement. So when Theodore Roosevelt appointed him the first Forest Service Chief in 1905, he eagerly adopted this European concept for all of America's many differing forested landscapes, landscapes he personally knew desperately little about.

......

The evolutionary reality of the western American forests is very different from that of northern Europe. For starters, North America possesses southern latitudes extending well below that of Europe. As a result, during the Pleistocene, while northern Europe evolved predominately Glacial type tree species, the North American west produced the two contrary tree types previously mentioned.

European style tree farming has proven itself ill-suited for the vast majority of the arid western landscape. The following are the primary reasons for its failure: slow growth rates, elevated fire risk and the still unrealized understanding that the tree type division must be restored and maintained.

There are however, other places where Cotta's method works quite well in northern Europe and west of the Cascades (Pacific Northwest). These successes can both be directly attributed to high annual rainfall. This abundance of water promotes accelerated growth rates, while providing elevated fire protection. These places are also predominately occupied by Glacial trees and therefore don't require tree type separation.

It must be remembered that our modern society requires vast

quantities of wood fiber to function. Tree farms, if located properly, are an environmental positive. Every tree grown within one of these man-made entities reduces the need to enter a virgin ecosystem somewhere else on the planet.

……

In 1905, the same circumstances that caused Europe to adopt Cotta's plan almost a century earlier also existed in America. Many of America's ancient forests lay in ruin. European style tree farming went on to replace some of those decimated forests.

However, in Northwest Montana because of the extreme terrain and limited infrastructure, full implementation didn't arrive until after World War II. This sudden widespread application was expedited by an exploding housing market.

This immediate expansion was made possible by the large supply of war surplus vehicles, which included trucks and dozers. New roads were soon being pushed into remote areas. The federal government and privately owned railroad companies were soon making clearcuts on their perspective ownerships. The removal of some of the last remaining Native American-influenced forest structures was soon well underway.

The massive upswing in timber harvesting shattered the former logging paradigm, where previously only the well-funded railroad logging crews had access to the latest equipment and best ground. Now anyone (gyppo) could start up a privately-owned logging company, provided they had enough money and/or bank backing to purchase a large performance bond and the necessary logging equipment. Once these hurdles were crossed however, the gyppo logger was free to bid on federal timber sales and get in on the good timber. Both company types (gyppo and railroad) were under strict supervision from university-educated foresters, who were tasked with installing this European concept.

Money still separated the two logging cultures, which unintentionally created a two-tier justice system with different consequences for the same offence. For example: if an employee working for a railroad crew either intentionally or by accident, deviated from the company foresters' instructions, the guilty employee would very possibly lose his job. The company crew however, as a whole remained unaffected.

If however, an employee working for a gyppo logger ran afoul of a federal forester, the gyppo logging company could lose its federal contract and forfeit its ability to bid on future timber sales, which would very likely force the company into bankruptcy.

While this powerful control mechanism did give the federal foresters the necessary leverage over the gyppo logging companies and thus enabled them to enforce the proper installation of European style tree farming, it unintentionally created a subtle negative side affect. The trial and error required to think outside the European tree farming box was no longer possible. Neither the large gyppo logger or federal forester would foolishly risk their livelihoods by purposefully departing from this government-imposed doctrine.

SMALL TIMBER GYPPO

At the time, virtually all individual loggers regardless of company type were paid the gyppo wage. The more timber you cut, the bigger the payday. This encouraged everyday loggers to be in constant pursuit of a more efficient way of cutting timber.

A stunning breakthrough came from a logger by the name of Joseph Buford Cox. By 1947, Joe had been using powered chainsaws for quite some time, but the existing cutting chains were slow and required constant filing. He took it upon himself to find a better way. He found the answer in nature.

One day while watching a timber beetle larva easily chew its way though wood, he became curious as to how it was doing it. He took it home and placed it under a microscope, where he observed opposing C-shaped jaws. This inventor with a fifth grade education went into his workshop and began experimenting. Using steel, he replicated the jaws and affixed them to a chain. A period of trial and error ensued, which soon led to success. His chipper chain dramatically increased cutting speeds and is still being used today.

I mention Joe's invention because it was the added speed and efficiency of this new saw chain that ultimately allowed Granddad, Ora and all the other families to give it a try in the second growth small tree forests. I will refer to these dreamers as 'small timber gyppos', which included aspects of both sawmilling and logging.

As stated in Chapter 3, this second growth forest was considered valueless by the big logging companies, not only because most trees were too small to make railroad ties (6"×8"), but also because most Glacial trees are structurally unsuitable for this purpose. The Glacial trees (inland Douglas Fir and others) are however, ideal for creating small dimension lumber found within home construction (2"×4"-2"×6"), which was suddenly in huge demand.

The land under many of these diminutive forests had been 'given' to homesteaders decades earlier (The railroad had often already removed the old growth). A high percentage of these families went on to operate small family ranches. Over the years, they'd watched the grass slowly decline as the second growth Glacial forest grew thick and shaded the forest floor. This reduced the number of cattle their limited land could support. They desperately wanted the once thick grass restored. These ranchers were therefore very receptive to the small timber gyppo and eager to give them a chance.

……

The sawmilling aspect of the small timber gyppo was just a scaled down version of the large headrig sawmill being used at Somers. To understand how a headrig works, one needs only imagine a short railroad with a single flatcar riding the rails. The log is placed on the flatcar (carriage) with the log hanging over the side a little. As the carriage travels forward, it just misses a large circular saw blade, which subsequently cuts off a slab from the overhanging log. The carriage is retracted. The log (cant) is flipped, adjusted and run through again. The process is repeated until the desired dimensional lumber is attained.

The logging portion started out as just a micro version of existing methods. Trees were cut down, segmented into logs, pulled to a centralized location (landing) and loaded onto a truck. The logs were hauled to the headrig sawmill and turned into lumber.

There were however, a couple of major differences between the big companies and the small timber gyppos. Firstly, there weren't any foresters involved in the beginning. Secondly, they were selective cutting and not clearcutting.

The phrase 'selective cutting' has a nice ring to it, but it's really nothing more than a variant of European style tree farming, which likewise myopically focuses on accelerating tree growth. This benevolent sounding method doesn't produce fire-resistant forest structure. To be fair, it does burn with far less intensity when compared to that of an unmanaged transitioned forest.

......

In order to achieve maximum efficiency, the small timber gyppos needed two crews, one in the woods and another at the mill.

Startup money was short for most and going to the bank wasn't an option, as these institutions rarely loan money to poor people touting untested ideas, especially when the big well-established logging companies believed it impossible to turn a profit in such small timber.

Undeterred, Granddad and Ora pushed forward, splitting the operation in two. The Bruches (Granddad and two of his younger brothers, Harold and Roy) would supply the logs, while the Henrys (Ora

Uncle Ora and Aunt Anna's three sons. This Photo was taken during World War 2, when Mel and Vern were back in Kila on leave (both served in the Army air Corp). Elwood soon joined the Marine Corps.
Photo provided by Carrie Henry-Ramsey (Granddaughter of Melvin)

and his three sons, Melvin, Elwood and Vernon) built and operated the sawmill, each family scrapping together whatever was needed in order to perform their particular aspect of the operation.

Granddad bought the chainsaw. He operated it with backward pressure from his right forearm hanging over the handle bar, while his left hand operated the throttle. Harold was a World War II veteran (He was with Patton during the Battle of the Bulge when they relieved the

101st Airborne Division at Bastogne) and therefore qualified for a free war surplus truck.

They still needed a way to load the small logs onto the back of the truck. This was accomplished with the aid of a Ford Model-A rear-end. A home-made cable drum was attached to one end, which successfully repurposed it into a winch. The cable ran up and over a boom they'd made, to a set of tongs. With sharp inverted points, the tongs bit into and held the log as it was lifted onto the truck. Money wasn't available for a dozer, so they used horses to skid the logs to the landing. Ora located the milling site on a flat spot uphill from Kila and just south of Highway 2 below Mile Marker 111.

They bought the sawblade, but almost everything else came from scrap iron and wrecked cars. They used a Model-T transmission to rotate a pipe that ran across and under the mill frame rails. A cable was triple wrapped around the pipe, with the opposing ends of the steel rope going through pulleys on opposite ends of the mill and affixed to the opposing ends of the carriage. The forward movement of the carriage was activated by a lever, towards the sawblade. This would prove a fatal flaw.

The expense of getting started had left Granddad and Grandma Bruch almost penniless. When Christmas of 1949 arrived, Grandma apologized as she handed Dad one present. He opened it and found a toy worth perhaps 10 cents inside. Dad knew they'd been struggling for money. After he unwrapped the gift, Grandma once again said she was sorry, explaining "there simply wasn't any money." Dad sincerely told her that he was very happy with his present and hadn't expected any. Her formerly forced smile broke and gave way to tears of joy. To his dying day, it was this Christmas moment that Dad most fondly recalled.

In the summer of 1955, Granddad offered my 16-year-old dad a job in the woods, which he happily accepted. On the first morning before they left for work, Grandma Barbara pulled him aside and out of earshot of Granddad, in a hushed but serious voice, told him not to embarrass his father by not working hard.

Arriving at the harvest site, Dad discovered one of his jobs was helping load the truck with Roy and Harold. Roy ran the winch, while Harold tonged the logs out of the deck. Dad was on the back of the truck, situating the upcoming log and freeing the tongs.

As they were riding home at the end of the day and seeking affirmation for a job well done, Dad asked Harold how he'd done? "Well," Harold replied, "You might make a good logger some day, if you weren't so damned lazy." It cut Dad to the core. He spent the rest of the summer at a run, trying to prove he wasn't worthless.

Almost everyone that has worked for their family's small gyppo operation has similar stories. The very existence of the small timber gyppo demanded they worked harder than their old growth counterparts and usually for considerably less pay. As a result, the initiation for the young aspiring gyppos was always a harsh awakening.

It all came down to size. Where one old growth tree in northwest Montana frequently possessed perhaps 3000 board feet of lumber (a board foot is 144 cubic inches). A second growth Douglas fir may only have a 100 board feet and often much less. To make up the difference, the small timber gyppo had to work at a pace akin to that of killing snakes. All the while seeking an innovative breakthrough in a never ending effort to close the profitability gap.

That breakthrough came in the late 50s. It seems so obvious now, but like all innovations it was invisible to everyone until someone stumbled across the solution. I refer to the Scragg Mill. It was simply two saw blades on one shaft, but it eliminated the carriage. Now, instead of retracting the carriage and repositioning the log during a series of several repetitions, small logs could be sent through the mill end for end, with two sides simultaneously rendered flat. Small log milling and logging innovations from this point forward took off like wildfire. Unfortunately, this giant leap forward came too late for my family.

High School photo of Melvin Henry
Photo provided by Melvin's Granddaughter, Carrie Henry-Ramsey

In 1956, the Henry mill, like everyone else's, was still

This photo of Melvin and his three children (Doug, Katie and Kellie) was taken a couple months before the accident.
Photo provided by Carrie Henry-Ramsey, daughter of Doug and Sue Henry

using the carriage. Melvin was the head sawyer. The pace of milling was brisk. Load a small log onto the carriage, run it through the blade, retract, flip, adjust, run it through, retract, flip, adjust and so on. Load the next log and again and again and again as fast as possible. His last log snagged his pants and began pulling him towards the saw blade. He frantically tried to push the lever away from himself, in a desperate attempt to reverse the carriage, but because it was already in the forward position, he didn't have the leverage. Out of respect for my cousins, I will not describe the subsequent horrific scene, but I'm sure your imagination has already painted the picture.

In the aftermath of Melvin's death, the determination to continue working with the small timber very nearly came to an end for the Bruch and Henry families, but eventually they decided to proceed.

The following year of 1957 saw Bruch Brothers logging finally replace the horses with a small John Deere dozer, but they decided against purchasing a canopy, which had the inhibitive cost of an additional $500.00.

Transitioning from horse skidding into using a dozer was a painful experience. The first week they owned it, Harold was in the operator's seat heading for the landing, when a log he was pulling hit a stump. The stump acted like a fulcrum, which pivoted the log up and over, and because there wasn't a canopy, it slammed down hard on the top of his head. Remarkably, it didn't kill him. With this painful lesson learned, no similar incidences occurred during the remainder of the summer and

This is the model of dozer Bruch Brothers Logging purchased (1957 John Deere 420). Internet photo

on into early fall.

Harold was Dad's favorite uncle (he had many uncles). Dad often reminisced about an incident that happened during his first summer of logging. Dad had been on top of the load when he slipped and fell, getting caught by the tongs in the process. He dangled precariously upside down, yelling for someone to help him. Harold was a ways off, but saw what had happened. He ran as fast as he could to get Dad off the sharp points, which Harold was sure had driven deep into his leg. When he arrived however, he found the tongs had harmlessly snagged his pant leg. Relieved and not willing to pass up such a golden opportunity, he swung Dad back and forth as he roared with good-natured laugher.

Fast forward two years to October 1957. The long work day was finally coming to an end. The perfect weather had encouraged Harold's wife Nancy to drive up to the logging site and give him a ride home. Granddad was having a conversation with her, when the noise of the dozer caught their attention. They looked up on the hillside as Harold crested a rise as he brought down the final 'turn' of logs for the day. The logs trailing behind the machine soon emerged over the lip of the hill. The log ends closest to the crawler began rising high into the air, maintaining the previous flat trajectory from which they were emerging. The logs' center of gravity changed halfway over the lip and came violently crashing down, creating downhill momentum. The logs slammed into the back

Uncle Harold in uniform. Taken during World War 2. Shortly thereafter, he left for the European theater.

of the little machine with a sudden jolt, pushing and sliding the dozer sideways. The downhill track caught a stump, which immediately began tipping the dozer over. Harold jumped off and tried to outrun the impending wreck, and for a brief moment it appeared that he might win the race but then he slipped and fell. They lost sight of him just as the dozer overtook him.

Granddad told Nancy to wait while he went up and checked on his brother. Harold had survived a Stuka dive bombing attack and countless other close calls during the war. He'd also taking part in the liberation of the Dachau concentration camp. Tragically, on a waning but still beautiful fall day, Harold Bruch had accidently found his final fate on that dry Montana slope. Granddad spared everyone the description of what he found at the conclusion of his grim journey.

In the aftermath of Harold's death, both the Henry and Bruch families decided it was time to end our involvement within the innovative small timber gyppo movement.

......

Other small timber gyppo families went on to shatter milling paradigms. The Larson family would push it further than anyone by developing the fastest mill of its kind in the world, all while cutting logs that were frequently far smaller than 12 inches in diameter. Their American Timber Mill was located along the banks of the Stillwater River, the same river Great-Grandpa Moses chased logs down near

Olney, Montana. At its height in the mid-1980s, if memory serves me, the mill could produce 385,000 board feet of small dimension lumber in a single 8-hour shift. To put that number into perspective, if 2"×4"s were placed end for end, they would reach 110 miles.

The small timber milling breakthrough had finally made small trees very profitable to harvest. By 1990 small tree logging technology had finally caught up with the evolution of feller bunchers, grapple skidders and log processors, which likewise didn't require large old growth trees to maintain profitability. It had been decade after decade of one giant leap forward after another.

The future seemed bright for the many small timber gyppo families who suddenly found themselves controlling all the timber harvesting and a high percentage of the milling here in Northwest Montana. That hope however, was soon replaced by a never-ending nightmare.

Speaking from hindsight, if we had been allowed the opportunity to continue down the innovative path with the freedom of trial and error, it's quite possible that the European tree farming doctrine, so ill-suited for this region, could have been replaced with a breakaway forestry method. It could have been a method more aligned with the successful outcomes produced by the Native Americans - a forestry method rooted deeply within the evolutionary realities of the Pleistocene. But it wasn't to be.

With the best of intentions, the environmentalists began utilizing the judiciary to appeal and ultimately stop virtually all timber harvesting occurring on federally owned land, which represents the vast majority of ownership in Montana. In an attempt to compromise with the environmentalists, Montana passed some of the most stringent environmental laws in the nation. These much-needed laws focused on protecting streams and other waters (streamside management zone). The timber industry also developed several other mitigating measures, which focused on minimizing impacts associated with logging (better management practices). Silt entry into streams from logging became a rare occurance. Clearcutting was also abandoned and replaced with selective cutting. These efforts were spearheaded by the large industrial timberland owners (Plum Creek and Stoltze), who eagerly demonstrated a willingness to change, in hopes the federal lands could be reopened.

The courtroom assault however, only picked up speed. This happened not out of environmentalism, but because the judiciary permited the misuse of the 'Equal Access to Justice Act'. Suddenly lawyers and environmental groups had a new revenue stream, which led to conflict profiteering.

The subsequent legal feeding frenzy quickly expanded to include any private or religious activities occurring on federal land (Ten Commandment monuments). The federal government (taxpayer) paid out billions. Many environmental groups became nothing more than shell organizations occupied by lawyers seeking wealth. They became completely uninterested in finding solutions.

……

As the federal timber started drying up, the gyppo loggers and the dozens of highly efficient family-owned mills began going out of business at an alarming rate. This eventually included the Larson family's American Timber sawmill near Olney.

The lumber has to come from somewhere. So with the noose suddenly tightened around the neck of the northwest timber industry and because the exploding world population still wanted to live in houses, the cut immediately moved to the virgin old growth forests of Canada and Russia, thus fulfilling Dad's warning from Chapter 2.

Strangely (tongue in cheek), no tariffs were demanded by the environmentalists, nor protests organized at our northern border to stop the ecological destruction. Cheap lumber soon began pouring into America, expedited in 1993 by NAFTA (North American Free Trade Agreement), which was nothing more than an environmental dodge. The destruction of these last untouched bastions of nature was accelerated by the very innovations devised by the American gyppo, which were now used against us. Innovation and manufacturing of this type of equipment moved both north and overseas.

Once again my family, along with the many other descendants of the original homesteaders, found ourselves directly paying the price for poor political decisions. These were being made by far away politicians, who themselves remained unaffected by the subsequent negative outcomes.

Suddenly, Plum Creek and Stoltze were about the only games around. Gyppo loggers who had formerly been harvesting federal timber began swarming to Plum Creek, with hopes of cutting on the million-plus acres of industrial timberland.

This lifeboat proved an unstable vessel for the multitudes seeking safety from the shipwreck that had become the Northwest timber industry. A bidding war broke out, with families competing for the few available timber sales. There wasn't enough to go around. The price of timber harvesting plummeted. Accident rates soared. Logging family after logging family began declaring bankruptcy.

Those that struggled on hoped against hope that they'd be able to weather the environmental storm. They steeled themselves with what seemed like a reasonable thought: they only needed to cling on just a little longer until the tempest subsided. It never did.

Today, almost all the gyppo logging and milling families are gone. Plum Creek, weakened by unfair advantages given to importers, also succumbed. In 2015, Plum Creek merged with Weyerhaeuser, who, four years later sold its Montana holdings to Southern Pine Plantations with the transfer completed in the spring of 2020, thus ending my family's long involvement with harvesting timber on this large industrial ownership. In the end, Bruch Brothers Line Logging was the last surviving yarder logging company working the company ground.

At its height, Plum Creek owned 1.6 million acres in Montana. Over the years it dwindled down to about 600,000 acres. Weyerhaeuser sold Montana because of its slow growth rates. Southern Pine Plantations has quietly been breaking up the acreage and over a third is already gone. The State of Montana needs to immediately buy back what remains. This land needs to have its forests restored and not turned into sprawling home developments.

Of the handful of us that somehow managed to survive this endless war of attrition, few encouraged their children to follow in their footsteps. As a result, every hard-earned lesson that has led to an unparalleled intimate understanding of the forests of northwest Montana is now at risk of being lost forever.

I can't speak for anyone else as to why they decided to stay in this dying American industry. As for me, it happened during a singular moment in time, which started me down the path in search of a solution. This event occurred in the late 80s.

At the time, I was killing hundreds of old growth trees each week. It was in the midst of this slaughter that I first became aware of the importance of the long gone Native American burning practices. This rare knowledge had come from another old growth sawyer, my great-grandfather Everett who I mentioned in chapter 1.

Up until that moment in 1986, I had falsely assumed the forests of Montana were natural and free from any manipulations of man. Suddenly,

I found myself completely reversing my way of thinking, which led to my founding epiphany that the Native Americans had developed a way to maintain the fire-resistant forest structure.

It was a pivotal moment. A rudimentary idea began formulating in my mind. If we managed the understory, we could allow the old growth to live out their entire lifespans in a manner reminiscent of the Native Americans, while at the same time restoring fire resistance.

Again, it was just a basic idea, which was just riddled and ripe with obstacles. Foremost was one enormous road block. While it was true that the small tree technology had drastically reduced size requirements, the ultra small size barrier remained and must be breached if my envisioned departure from current methods had any hope at success.

……

In the aftermath of discovering this new found knowledge, I was gripped with a terrifying vision, which allowed me to accurately see what the future held. All the fire-resistant old growth trees (Inter-glacial) were going to die, regardless of whether we logged them or not. Unstoppable massive canopy fires would become commonplace. These canopy fires would burn so hot that the seed sources would be consumed along with the top soil. This would not only spell the end for many of our western forests, but quite possibly lead to complete ecological collapse. Again, at the time, I was completely unaware of the climate implications.

Today, sweeping mega fires have indeed become normal, along with a massive die-off. The unstoppable fires are rapidly consuming huge tracts of arid 'out of phase' western forests, which wrongly believe themselves growing during a wet glacial cycle.

Weather patterns have begun changing, which is not only heavily influenced by our own transitioned forests, but also by the massive deforestation that have been occurring in Canada and Russia during the last thirty years. This massive clearcutting correlates with the permafrost beginning to melt in the late 90s. The mega fires in these northern latitudes that were once rare have become disturbingly commonplace.

This all points to one terrifying apocalyptic conclusion. We are witnessing the reemergence of the long absent Altithermal.

……

After Harold's death, Grandad quit logging and focused on his few head of cattle. Grandma Barbara encouraged Dad to do anything except log. She offered to get a job for the explicit purpose of paying for his college tuition. He took the offer seriously and thought of perhaps becoming a lawyer (you wouldn't be reading this book, if he'd gone down that path).

Dad ultimately decided to stay in the Flathead, first working at a tire shop, then shortly after getting married, for a large plywood mill in Evergreen. Roy passed away from diabetes in the early 60s, making Granddad the last surviving member of Bruch Brothers logging.

......

While still in high school, Dad joined the Army National Guard in 1956. On May 9 in 1965, it was drill weekend. He was a radioman, in the process of setting up an antenna in Depot Park in downtown Kalispell with his brother-in-law Larry (Mom's brother).

The dense cloud cover broke a little, letting in a stream of light. Dad pointed at the cross, which had been created by the antenna's shadow. He said half jokingly to Larry, "That'll make you religious." Within seconds, an officer approached and said, "There's been an accident."

Each year, several small ranchers got together for the purpose of helping each other brand the new crop of calves. It had become a social event, with wives and husbands finally getting a chance to interact with others after the long cold winter.

Granddad's prowess at riding and roping were well-established. He was in the midst of applying this talent, when Grandma Barbara heard the sound of metal on metal approaching from behind. She turned and saw a 3-year-old child playing in the road. She had no more then a few seconds to react. The young girl's life hung in the balance.

A few minutes prior, a fully loaded fuel truck had backed passed her as it carefully navigated between the many cars parked along both sides of the road. The incline of the road steepened as the trucker slowly climbed the hill towards the awaiting fuel tanks. The fuel truck was brand new and equipped with a 'cutting edge' type parking brake.

The trucker finally reached his destination, pumped the brakes and flipped the 'micro-lock'. He got out, unreeled his fuel hose, climbed the ladder to the top of the ranch's fuel tanks and began filling. Some time had passed when the trucker, who was focusing on the task at hand, suddenly had the nozzle ripped from his grasp. He looked up in horror as the truck began rolling forward. It quickly gained speed and momentum as it began bouncing off parked cars, which concentrated and maintained its deadly trajectory. He jumped off the tank and chased after the truck. It would prove an unwinnable race.

The sound of grinding metal caused Grandma Bruch to turn. The two women with whom she'd been conversing described what happened next. Without hesitation she raced towards the child, grabbed her and threw her under a parked car. A split second later the truck struck Grandma, killing her instantly. The girl survived unharmed.

Yes, my Grandmother was a true hero. Others thought so as well. She was posthumously given the Carnegie Award (civilian version of the Medal of Honor). Her selfless courageous 'decision' isn't why I'm reliving this tragic event, although she very much deserves to be remembered. I'm revisiting this incident in order to discuss the decisions Grandad made in the aftermath of the accident. It was his subsequent example that caused me to rethink my life following the 1990 logging accident, after which I stopped pursuing arbitrary societal stratification.

The fuel truck driver was devastated. A few years ago, I met his daughter for the first time. She revealed to me that her Dad never recovered. The accident continued to haunt him until his dying day. I told her my family had always felt nothing but compassion for him.

This family position had been nurtured by Granddad. In the aftermath of the accident, his primary concern was for the inconsolable trucker. No stranger to accidents, Granddad must have had a fairly good idea of what the trucker was going through.

A few days later, Granddad asked the trucker to be a pallbearer. Ugly talk soon began circulating, with a few people whispering into Granddad's ear, saying things like, "Nobody in your family will ever have to work again." Granddad would erupt and say he wouldn't take a dime for his wife's death. It was blood money and nobody was going to place a dollar amount on her life. There will be no lawsuit!

The insurance company kept pestering him, trying to get him to take something. He would run them off his property. Finally, an insurance agent was able to convey to him that the books couldn't be closed without some kind of payout. Not wanting to make the agent's life harder, Granddad agreed to a minimal payout that he promptly gave away, splitting it between Aunt Virginia and Dad.

I was well aware of my grandparents 'decisions', but I never really fully understood until after the 1990 accident just how important it was to place monetary reward behind doing what's right. This includes refusing to accept tainted money.

Our individual collective 'decisions' will echo for an eternity.

......

About a year after the Grandma's accident, Dad decided to become an old growth sawyer. He came to Granddad for some instruction. Granddad refused.

Dad, however, was determined to do it with or without him. Granddad relented, out of fear his son would get himself killed and agreed to tutor him for a few days. Dad soon learned the basics of how to safely cut old growth timber, doing so while employed by a large gyppo logging company.

On that first day of instruction as they drove along the brush piles, Granddad marveled at the waste. He told Dad that the tops being thrown away were bigger then the trees he and his brothers had been cutting (minimum top size at that time was 8"). Today, because of the small timber gyppo, top sizes as small as 2" are marketable.

Logging within the second growth forest in the 1970s, Dad proudly stands next to his 1st generation rubber tired skidder (Garrett), while holding his Homelite chainsaw. A happy time.

……

Dad sawed the old growth for a few years. Then in the early 70s, Dad and Granddad began working together, once again immersed within the privately owned second growth forests. First generation rubber-tired skidders and smaller/better chainsaws finally made it somewhat profitable.

It was during this era that Granddad revealed much of what you've been reading. Sometime during this timeframe, Dad came to a realization. The relationship between him and his father had transcended into a close friendship. On one of those days, while in deep introspective thought, Granddad said, "What happened to the Native Americans was wrong."

In early August of 1973, after what proved to be his last day of logging, Granddad went home…laid on the couch and fell asleep forever.

This image was captured in 1966 Kila, Montana. Granddad always carried himself with dignity and never complained about the misfortunes and hardships that marked his life. His face in this grainy photo however, reveals an unmistakable sadness. Undoubtedly, the product of the multiple tragedies, which'd marred the previous 10 years.

CHAPTER 5
THE BRUCH THEOREM

Dad's (Marlowe Bruch) High School photo.

Located in Northwest Montana, Smith Valley is a somewhat small glaciated valley connected to the vastly larger Flathead Valley from the west. Long ago, as a several-thousand-foot thick glacier slid south down the Flathead Valley, a small finger glacier branched off from the main flow and drove west across the northern half of Smith Valley, terminating a couple of miles later.

Today, the path of that long gone Smith Valley glacier can be traced by the lateral moraine (a ridge of material deposited along the edge of a moving glacier) it left behind. In 1960, my newly-wed parents bought thirty acres of that ancient lateral moraine from Aunt Violet and built a house at its base.

At that time, this acreage was considered low value because the rounded treeless ridge was ill-shaped for either a wheat or hay field. As a result, everything on top remained undisturbed by modern man and therefore still retained its native bluebunch wheatgrass.

It wasn't the grass however, that initially caught my attention as I

Mom's (Frances Carpenter-Bruch) High school photo.

explored the sparsely populated landscape of my youth, but a large rock that I'd found on top of the ridge. This boulder was different from the 'roundish' rocks occupying the gravel pits flanking my parent's property. This one was angular with somewhat sharp edges… why?

The shape of the roundish rocks was easy to explain. They'd been rolled by a glacier, which had broken off the angular edges (glaciated). Those flanking gravel pits were exclusively occupied by glaciated rocks of various sizes and colors (many having been transported from Canada)

Dad digging a hole on his and Mom's new property. The lateral glacial moraine can clearly be seen in the background. The ridgetop boulder (not seen in this photo) is located on top and further to the right.

and interspersed with the powdery remainder of pulverized stone (rock flour). That anomalous unglaciated ridgetop boulder however, just stood there defying explanation, taunting me to unravel the mystery of its placement.

Ridgetop boulder

It would take most of my life, with clues coming in at a glacial pace, but eventually, the circumstances of how that rock came to be situated on top of that ice age moraine became clear. That angular rock is an evidentiary link connecting a chain of events that culminated in the release of the enormous Lake Missoula flood 14,800 years ago. The unusual location and shape of this rock provides a physical clue into what happened just before and during that cataclysmic outburst.

Current scientific dogma tells us that Glacial Lake Missoula resided dozens of miles south of Smith Valley, its northern waters hemmed in by the afore-mentioned main glacier, which terminated just south of today's Flathead Lake. Another glacier to the west (Sandpoint, Idaho) was functioning as the cork in the bottle. It's widely accepted that this ice dam broke and unleashed a humongous flood, which swept across portions of what today is Idaho and Washington State. The torrent profoundly altered the downstream landscape (channeled scablands).

After the lake emptied, it's believed the Sandpoint glacier, which was sliding down the Purcell Trench, once again blocked the outlet, causing Glacial Lake Missoula to refill - only to once again collapse 55 years later and drain the lake. This process was repeated over and over. All told, it's currently believed there were perhaps 40 or so fillings, spanning a period of about 2000 years.

These numbers were initially generated by counting the ancient lake shores found on the hillside above the U of M campus in Missoula, Montana. Its assumed that every lake shore is representative of a 'filling', with each consecutive lake making a new lakeshore just a little below the former lake level. A guess of about 55 years per filling was attached to the theory. Common sense just screams that something is a little too consistent with this multi-filling theory.

David Alt, who wrote the book 'Glacial Lake Missoula', also believes something is amiss. His skepticism fueled by several unexplained water-eroded ravines that cut down through all the lake shores. Some of these 'wash outs' begin well above the highest horizontal line, yet inexplicably lack any ground-fed water sources, giving rise to a couple questions: under what circumstances were these ravines formed and are they associated with the Lake Missoula event?

Missoula lake shores, located above the U of M Campus.
Photo by Don Hyndman, courtesy of the University of Montana
Missoula, Montana

I contend the unique placement of that Smith Valley ridgetop boulder provides a powerful clue into how these 'lake shores' and eroded ravines came into existence. It began when a massive water source came slamming down from the North. This expansive flood likely overwhelmed and triggered the collapse of the Sandpoint ice dam, while also creating a strong circular current within the fairly round Missoula Valley. As the turbulent lake drained, the hillside lines were carved (bathtub effect). The tail end of a lengthy and extremely powerful rainstorm (perhaps lasting 40 days and nights) pelted the empty lake bed and surrounding landscape, flushing the ravines into existence.

Evidence of this mysterious northern water can be found sporadically throughout Northwest Montana. The most compelling is located at the confluence of Smith and Flathead Valleys. Driving west on Hwy 2 and taking a right just after mile marker 117 will place you on West Valley Drive. A quarter mile later and just before an abrupt right corner, you are provided with a direct view of where that ancient glacial finger initially branched off from the main flow.

Today, the residual glacial debris forms a plateau about 60' high that juts up about 150 yards from the road corner. You will immediately notice the mouth of a valley projecting backward through the glacial deposits. According to current paradigms, this valley shouldn't exist because this area was supposedly under several thousand feet of ice during the entire existence of Lake Missoula. Yet, there it is…

Vortex valley, Google Earth coordinates (48°11'39'N 114°23'58'W)

I walked this anomalous valley countless times during my youth. There isn't a creek bed on the valley floor, which dismisses the possiblity that this unexplained trench about 60' deep and 125' wide at its mouth was formed through the process of long-term water erosion. When I reached the end of this slowly curving half-mile long valley during my youthful expeditions, I was met by a chaotic collection of angular boulders, which had been dislodged from their local bedrock moorings.

Walking up its length, the valley slowly loses both width and depth as it gains elevation on the ridge that separates Smith from West Valley. The ridgetop point of this now necked-down valley (perhaps 20' deep'×

30' wide), cuts down through hard granite (48°12'08'N 114°24'14'W). It's readily apparent that a large portion of the stone is missing, having been ripped loose and transported elsewhere. The previously mentioned angular ridgetop boulder very obviously originated from this spot and became summarily deposited three-quarters of a mile to the south.

It wasn't alone, as several more examples of angular-shaped rocks are sprayed out across the landscape. Collectively, they all came to rest in an arc-shaped pattern. This semi-circle arrangement starts to the southwest of the valley, and from there the strewn pattern runs easterly, with several examples located a couple hundred yards out in front of the valley. The final leg of the arc wraps to the northeast, along West Valley Drive.

This angular boulder debris field is an undeniable indicator that an underwater vortex had formed at this location as Lake Missoula drained. This aquatic tornado was ushered into existence at the intersection of two extraordinarily strong water currents (one flowing south and the other east). This powerful force ripped the boulders free from the bedrock and carried them southward as the whirling waters quickly carved through the glacial debris field, forming the valley as it went. The rocks dropped into their long term positions when the vortex abruptly subsided.

The existence of this excavated valley gives us another clue. At the time of its formation, the covering glacier at this particular location was either very thin or completely absent.

Unfortunately, because of robust home development in the recent past, the orientation of several of these boulders have been compromised. Many however, are resting exactly as I remember. The ridgetop boulder remains undisturbed. Another several hundred ton example is located on Morning View Drive, about 200 yards west from where this road branches off from West Valley Drive. This sleeping giant is partially hidden in a hedge, about 150' north of the road (48°11'25N 114°23'55'W).

So to recap, rocks were torn from the bedrock, spewed south and east by the actions of an underwater vortex, which simultaneously excavated a valley. For most of my life, this was the extent of my incomplete theory, which without a water source was unviable…

And then I found it…

A few years back, while in Canada searching for a piece of used logging equipment, I drove through Canal Flats (50°08'50'N 115°48'12'W). A siren suddenly erupted in my head. This area is just south of Columbia Lake (origin of Columbia River) and just north of the Kootenai River. Columbia River exits the lake to the north while the Kootenai flows south into Montana. It wasn't the directions these rivers were moving or the very close proximity to one another that initially caught my attention. It was the wall of rock flour, dozens and dozens of feet high right next to Highway 93, which indicates the presence of a long gone glacial lake. The rock flour was once at the bottom of that lake and the sheer volume of pulverized stone meant this former glacial lake had once dominated the area for a very lengthy period of time (the rock flour settled onto the lake floor, after being transported to the former glacial lake by streams).

Something was missing however. If Columbia River had always drained north as it does today (eventually swinging south), a ridge composed of rock flour should still be located on top of Canal Flats... but it isn't there. A giant scar on the adjacent east mountain provides a clue into what happened.

At the end of the Pleistocene, an unprecedented melt event was underway, which was rapidly reducing the enormous glaciers and releasing tremendous stores of glacial melt water. Many inland mountain valleys soon began filling with water. The meltdown eventually reached a climax. The lake-retaining glacier that held the southern end of Pleistocene Lake Columbia, in residence long before the rapid melt down began, floated and broke. The ice dam fractured just behind the east mountain, which caused the ice to hinge southward and rip a chunk off the mountain as it went.

The opened flood gate unleashed an enormous southbound torrent. The flood waters carried massive quantities of lake bottom rock flour. The epic tide flowed down the Kootenai Valley and spread into adjoining valleys during its southbound journey. Today, sporadic placements of Canadian rock flour can be found here and there throughout Northwest Montana, giving us a glimpse into the massive magnitude of this flood. Although obscure, I'm not the first to suggest a southbound Lake Columbia flood is responsible for the erratically placed rock flour.

The following will focus on the path of evidence the flood left behind as it flowed, paused here and there, scoured and ripped its way through the western portions of Flathead Valley as it journeyed towards Missoula Valley. Since we've already discussed 'Vortex Valley', I'll exclude it.

The muddy flow that entered the Flathead Valley followed the path of today's Stillwater River (yes that river). At the time, a substantial glacier was splitting the northern half of the Flathead Valley down the center (today, it's an open-ended geological trough). The west and east lateral edges of where this former glacier was once located are very pronounced.

The western edge is locally known as Hillcrest, which is a several-mile-long abrupt elevation change (about 60'). The lower elevation is free of rock flour, indicating the glacier was still present and tall enough to prevent the flood waters from cresting over the top. The sediment-filled torrent slammed against this western facing ice wall, which constituted the southern projection of this waning glacier. The muddy pulverized stone formed a plateau west of the glacier. Today, Kalispell rests upon a thick layer of that Canadian rock flour.

Kalispell's eastern border is the Hillcrest. Driving east on US Hwy 2 down this hill will take you under a railroad overpass (entering Evergreen). Directly in front of you is an interesting hill (Conrad Cemetery). That long ago glacier must've had a hole in it, because this irregularly shaped hill owes some of its height, if not all, to a thick layer of rock flour.

A half mile south of Conrad Cemetery is a somewhat shallow terminal moraine, indicating there was an ice free zone beyond, which allowed the muddy water to spread out and cover a wide area of the lower Flathead Valley. A few miles later, the advance of the floodwaters were interrupted by a stretch of high stagnate ice, which was sitting atop today's Flathead Lake. This barrier redirected and intensified the southbound flow along the west side of the ice, ripping bedrock loose and forming craggy ravines as it went.

Political Hill is located on US 93, just south of the town of Lakeside. The roadway is cut deep through the hill, with the very center comprised of solid bed rock. On either side of this rocky partition are slopes filled with rock flour. The north slope contains rock flour mixed with some smallish rocks. The south side flour however, is interspersed with a substantial quantity of various sized boulders. This indicates the hill had disrupted the current enough to cause some of the heavier rocks to fall out of the flow and become deposited on the downstream side.

Further down the west side of the lake is Deep Bay. The floodwater apparently hit an abrupt edge of the melting glacier, because this portion of Flathead Lake is deeper than the surrounding lake bed. This suggests the interrupted fast-moving water excavated the lake floor, while simultaneously ripping up and removing bedrock from the surrounding area. Continuing a few more miles south, the torrent hit a westward bulge of the glacier and created Chief Cliff (47°51'12'N 114°19'22'W).

The floodwaters eventually arrived at the terminal moraine (Polson) and chose the path of least resistance by breaking through the western lateral moraine. The flow skirted around the western periphery of the massive ice/debris-covered terminal moraine.

Today's Flathead River occupies the path of that long ago torrent, and in places is running in a deep canyon of torn bedrock (Kerr Dam is secured to those rocky walls). Rock flour is stacked high atop this sturdy foundation.

The white cliffs and surrounding grass covered hills are composed entirely of Canadian rock flour. These hills are located above the banks of the Flathead river, near Kerr dam. There are few rocks found in the claylike substance, which indicates the formerly fast moving water had slowed after colliding with the main lake Missoula. The stalled flood water allowed the sediment to fall out and create these hills. Photo taken by Carrie Bruch

The flood waters rounded the terminal moraine and eventually breached through the last of the ice and collided with the deep waters of Lake Missoula. The strong current projected southward through Ravalli and on into Missoula Valley, where as previously described, it provided the energy for creating the 'lake shores'.

These however, aren't the only 'lake shores' the flood waters made. Another series of lines not far south of that long gone terminal ice, can be seen on the hillside south of St Ignatius (47°17'05'N 114°07'17'W). This soon to be 'lined' ridge was south of the ice and therefore already submerged under Lake Missoula when the flood waters breeched into the lake a few miles to the west. The culmination of the 'lined' ridge to the south, a mountain range to the east (Mission Mountains) and the terminal ice wall to the north, created a fairly round box canyon. When the raging waters broke through and crashed into the main Lake Missoula waters, it generated a turbulent circular current within this boxed canyon. This spinning water produced the horizontal lines as the lake lowered, replicating the process that was simultaneously forming the Missoula Valley 'lake shores'.

There are two more sets of hillsides lines. One set is located within the Niarada/Hot Springs Valley, while the other remains unidentified by science. The lines of the latter are very faint, but visible when conditions are just right with light intermittent snow cover. They are located on several mostly treeless hillsides a few miles north of where Flathead Lake empties into its namesake river. Driving north on US 93, one example of these overlooked multi-lines are located on a hillside northeast of mile marker 65.75.

Again, how is this possible if according to current scientific dogma, this particular area was still under extremely deep ice during all of the Lake Missoula episodes. These lines are situated exactly where they should be if flood waters came slamming down from the north and began swirling behind the massive terminal moraine. That water soon broke through the western lateral moraine and etched out these hillside lines.

The other three sets of 'identified lakeshore lines' were made on saturated hillsides. This saturation aided in the 'lake shore' excavation. The faintness of the 'unrecognized lines' strongly suggests this area was both ice free and above the water prior to the Canadian flood entry.

......

There have been several torturous attempts to study the layers of sediment found within the channeled scablands of Washington state, with some suggesting they've found evidence supporting the multi-flushing theory. This research has focused on slack water areas, where the water pooled up in valleys outside the main flow. There does indeed appear to be 40 nearly equal layers of sediment (rhythmites). However, these rhythmites require the same unrealistic consistency as those shorelines in Montana demand, with each successive flood depositing nearly identical amounts of sediment.

This seems highly unlikely, because the 'first' Lake Missoula flood event would have not only removed the lion's share of the loess (wind deposited soil) that formerly covered the scablands, but also created channels for the progression of weaker and weaker floods to follow. That would have greatly diminished the opportunity for the supposedly dozens of subsequent floods to pick up more soil.

One such slack water area is in Buringame Canyon in eastern Washington. Its 40 or so layers are interspersed with vertical fissures named clastic dikes. These vertical fractures run through almost all the horizontal layers, which further throws the multi-flushing theory into question.

The multi-flushing theory also requires the Purcell Trench glacier (Sandpoint) to remain unaffected by the 2000-year-long meltdown. The theory suggests that after each outburst, the glacier continued sliding south and dutifully blocked the outlet, which once again allowed the next Lake Missoula to fill. This is problematic because the Purcell glacier would need to continue growing if the glacier was to maintain both its size and movement. All the while, higher-elevation glaciers would need to be melting to supply the tremendous amounts of meltwater that was required in order to fill those 40 separate incarnations of Lake Missoula.

The multi-filling theory ignores other macro elephants in the room.

When driving in Canada along today's Columbia Lake, you are confronted with an overwhelming amount of rock flour, which stretches far above the contemporary lakeshore. The further you drive, the more awestruck you become by both the sheer volume and fairly even distribution of the rock flour, which is spread throughout the entire length of that ice age lakebed.

This consistent covering is completely absent on much of the landscape that was once submerged under Lake Missoula. This thoroughly contradicts the multi-filling theory, which if correct should have produced a somewhat thick and even layer of rock flour covering the entire lower lake levels. After all, with some short term interruptions, the lower portions of these 'lakes' were supposedly under water for almost the entire two thousand year period.

Instead of an even placement of rock flour, many areas are completely devoid of the substance, while other locations are piled high. This sporadic placement suggests the rock flour was injected into the Northwest Montana valleys from an outside source. This lack of consistent distribution also implies that Lake Missoula itself was both short lived and filled with fairly clean melt water, not the dirty waste water produced by the grinding actions of moving glaciers.

When this incident occurred, some or perhaps all of Northwest Montana's glaciers appear to have stopped moving, the Flathead flow having become two disconnected segments. The Canadian flood waters undoubtedly expedited the ice meltdown along their route.

The northern chunk may have still been moving, but since it only produced a shallow terminal moraine south of Conrad Cemetery in Kalispell, its final movement had been minor. The southern segment covering Flathead Lake was stagnate ice.

There's another problem with the current multi-filling theory.

When considering the enormity of Glacial Lake Missoula and the 40 different times the lake supposedly existed, it's reasonable to expect to find hundreds of miles of multi-layered shorelines. Yet, only four 'sets' of 'shore lines' have been located. The fact that multi 'shorelines' are only found in areas capable of producing rotating water, point to one conclusion…

Lake Missoula was a singular event and emptied just one time.

……

Lake Missoula wasn't the only enormous western glacial era lake that released massive amounts of water at the end of the Pleistocene. Lake Bonneville was another, and contrary to Lake Missoula's multi-flushing theory, it's agreed that Lake Bonneville drained just once. The current timeline has this lake emptying sometime around 14,500 years ago.

I challenge these differing release dates and propose a singular concurrent event.

Lake Bonneville was a huge lake covering most of today's Utah, with some of the lake also located within Idaho and Nevada. The old shorelines can still be seen high on mountain sides above Salt Lake City. It's currently believed that 14,500 years ago, the retaining dam holding back this giant lake, which consisted of overlapping alluvial fans (outwash sediment), failed and released the lake in a single huge flood event. The flood waters followed the Snake River Canyon and eventually reached the Columbia River corridor, which at the time was transporting the Lake Missoula flood. The waters merged together during this final leg towards the Pacific Ocean. (This last westerly stretch of the river is the border between Washington and Oregon).

Today, debris from the Bonneville flood can be found along the entire length of the Snake River Canyon. One hundred miles upstream from where this river meets the Columbia is a place called Tammany Bar. A gravel pit located within the outwash debris has revealed an interesting stratification. The top layer is composed of Lake Missoula silt, while gravel from the Bonneville flood occupies the lower. This is more than a powerful clue, it paints a clear picture.

As the muddy waters of Lake Missoula traveled south following the Columbia River Gorge in Washington State, the epic flood was suddenly forced to flow through a narrow gap (Wallula Gap). The Lake Missoula flood stalled as the water waited to squeeze through the necked-down corridor. The interrupted torrent backed up into several valleys, including the Snake River Valley. The fast-moving Bonneville flood debris slammed into this silt filled backwater, causing the heavier gravel to drop out and become deposited under the 'still suspended' silt. Tammany Bar's unique two-tier stratification provides us with undeniable confirmation that these two massive floods from Lake Missoula and Lake Bonneville happened concurrently.

The Smith Valley ridgetop angular rock and Tammany Bar tell us something else. During the last moments in which these giant lakes existed, an accelerated glacial melt event was taking place that was being driven by an extended period of heavy rainfall.

The Pleistocene ice dam located above present-day Canal Flats in British Columbia failed first. It was soon followed by the ice dam holding back Lake Missoula and finally Lake Bonneville, with each lake releasing massive stores of water. Lake Bonneville wasn't fed by glacial runoff, which is yet another indicator that an unprecedented rainfall event had taken place.

Wallula Gap, which stalled the flood waters, has almost twice the carrying capacity of the Clark Fork River gorge (exit channel for Lake Missoula). This means that Lake Missoula alone didn't produce enough water to back up at the gap…but lots of water did back up, supplying yet another layer of evidence that the Lake Bonneville flood was near synchronistic with the Lake Missoula outburst.

All this evidence leads to one conclusion. An extended and unprecedented heavy rainfall event had produced a region-wide rapid meltdown, which included the Canadian glaciers that terminated along the northern edge of Washington State.

A GLACIAL FOREST BURN OFF EVENT TOOK PLACE AT THE END OF EVERY GLACIAL CYCLE.

A cyclic 'Glacial tree' burn off event was already well underway during the lead up to the Lake Missoula/Bonneville flood event. This fiery occurrence was the product of Earth's strange orbital cycles. These include the changing eccentricity of Earth's orbit, which slowly transitions from more elliptical into less and back again (100,000 years). Earth has a wobble (26,000 years), which changes our planet's tilt over time, made wobblier by the oddly shorter procession of equinoxes (23,000 years). All these cycles combine to change how much solar energy northern latitudes receive. This greatly influenced floral and faunal evolution, along with plant community distributions. At the end of each preceding Pleistocene cycle, a widening latitude of 'fire-intolerant' Glacial forest found itself high and dry.

According to the Vostok ice core analysis, temperatures began rising at the end of each preceding glacial cycle. This slowly caused northern summers to get longer and hotter. Eventually (800-1000 years), lightning ignited the Glacial forest. These fires signaled an end to a very long period of elevated fire protection, which these now 'out of phase' forests had previously been afforded by the cold and wet of the prior glacial cycle.

Soil desiccation, which increases fire susceptibly, was intensified by the Glacial tree inability to slow transpiration. The Glacial forest, which never developed fire-resistant traits, exploded into all consuming canopy fires. This cyclic burn off event explains why CO_2 lagged behind temperature, which until now has been a mystery.

This cyclic Glacial forest burn off occurred at the end of every 100,000 year long glacial cycle. Smaller burn off episodes likely happened in-between. The conflagrations released vast quantities of CO_2 into the atmosphere. The Glacial forest itself very obviously emitted huge sums of CO_2 but it was the topsoil under the forest that held the lion's share of the stored carbon, which also ignited.

This low point in the glacial cycle was suffering from near carbon starvation. The sudden freeing of huge quantities of CO_2 and the rapid increase of thermal energy, which was reflecting off the burned landscape, combined to create a powerful greenhouse event. The elevated temperatures melted substantial portions of the glacial ice, which released huge quantities of melt water (perhaps accelerated by an accompanying rainfall event). The Grand Canyon was initially formed and subsequently deepened with each successive cyclic flushing.

The burned off Glacial forest was eventually replaced by an Inter-glacial 'fire-resistant' forest. Its fire resistance was substantially increased by the actions of the mastodon. Earth slowly descended back into the next ice age.

This unknown width of latitude west of the Rockies, which fluctuated between the two forest types, once again slowly ebbed back into a Glacial forest. Bushy/woody vegetation displaced the grass. The mega fauna grazers became replaced by smaller browsers. The large grass eaters were tethered to the fire-possible ecosystems, because this is where the thick lush grass was found.

The mastodons fight a slow losing battle against the increasingly wet landscape and the Glacial forest encroachment. The Inter-glacial forest retreated southward. The glacial cycle deepened. Atmospheric CO_2 slowly declined.

MAN ENTERS NORTH AMERICA FROM THE WEST

Living a divided existence, the two incompatible forest types had evolved in the absence of man. Nature was therefore defenseless when our 'intelligence' finally arrived. This happened just as the last glacial cycle was coming to a close.

I contend that a few bands of humans arrived on the west coast of North America and unintentionally intensified the effects of the cyclic Glacial forest burn off event, which was already underway. They did this by purposefully igniting canopy fires.

It's very likely they burned off both forest types, but I propose that the primary focus of these 'hunters' was within the Inter-glacial forests, which supported a rich diversity of animal life (mammoths, horses and many others).

These arsonists deployed fire as a very simplistic hunting tactic. It consisted of them arming themselves with long poles tipped with flame, which they used to reach up into the high forest canopy and ignite the highly flammable treetops. The strong, east-blowing prevailing winds did the rest.

This lazy but effective 'hunting' method allowed them to simply follow the path of devastation and discover what was for lunch. In so doing and because animals are deathly afraid of fire, they freed themselves from possible harassment from any large scavengers/predators, such as the short faced bear (extinct).

This unsavory scheme permitted them to leisurely stroll across the blackened landscape and pick and choose from a wide variety of victims that had been unable to evade the fast-moving overhead blaze. The intense heat had burned the lungs of these animals, which either killed them outright or rendered them unable to resist when the 'hunters' arrived for a meal.

This method likely didn't produce the same results within the Glacial forest, because it would have burned much hotter at ground level, degrading or destroying the food value of the smaller animals that resided within them. The eastern advance of these 'hunters' was eventually blocked by an obstacle (extreme terrain or body of water), at which point they simply walked back west and started the process anew on the next strip of unburned forest.

This anthropogenic burning dramatically increased the amount of exposed ground and atmospheric CO_2. This combined with the already substantial emissions being generated by the cyclic northern Glacial forest burn off, which magnified into an unprecedented greenhouse event (the vast majority of the topsoil under the Glacial forest was deposited on the landscape during the previous Inter-glacial forest phase). This ultimately produced more water than Pleistocene Lakes Columbia/Missoula/Bonneville were capable of storing.

I therefore contend it was the actions of just a few people that caused the Lakes Bonneville/Missoula/Columbia to reach critical mass and subsequently explode into a cataclysmic flood event (the oceans would eventually rise several hundred feet). The aftermath of the Lake Missoula flood saw the Pleistocene climate begin to fluctuate.

Three thousand years later, the Clovis people used the same hunting tactic on the Great Plains Inter-glacial forest. These makers of remarkably exquisite spear points (Clovis points) were perhaps the descendants of the original destroyers. It took them about 300 years, but the Clovis culture eventually removed the once enormous Great Plains Inter-glacial forest.

In the absence of thermal-absorbing and water-storing capabilities of the Great Plains Inter-glacial forest, the Pleistocene climate collapsed. This of course included the Great Plains ecosystem and food source it formerly provided, thus explaining why nearly all the mega fauna went extinct. The mega fauna had been the driving force behind the

Pleistocene glacial cycles.

The slow and steady decline of CO_2 that had taken place during those long gone glacial cycles was a direct result of atmospheric CO_2 transitioning into grass. The grass was consumed by grazers, which soon deposited their carbon-rich droppings onto the ground. This biocycle created deep, carbon-rich top soil (carbon sinks). This carbon sequestration mechanism was so effective that it reduced available atmospheric CO_2 down to 180 ppm by the end of each glacial cycle. This is near the point of carbon starvation (plants begin to die at 150 ppm). This CO_2 prison and subsequent release combined with the orbital anomalies to create a mirroring effect. This 'mirroring' is what is depicted on the sweeping ups and downs represented on the Vostok ice core analysis graphs.

The keystone specie was the mastodon, which maintained the separation between the two evolutionary opposite tree types. After its extinction, the ecosystems unraveled. The mirroring of orbital influences ended forever, explaining why the graph nearly flatlines during the ensuing Holocene.

It wasn't until the tree type separation was restored by the Native Americans beginning about 5000 years ago, that the climate began to stabilize and the rains returned.

This too was undone.

Modern fire suppression has unintentionally permitted a Glacial forest to slowly replace the western Inter-glacial forests. This ended the durable, although anemic remedy that the Native Americans had previously applied to the landscape with their former frequent burning. Those long-held traditions had been indispensable for maintaining the fire resistance of the western Inter-glacial forests during the last half of the Holocene.

The mega fires, the rekindling of a new mass extinction event, and the destabilization of the climate (beginning stages of a resurrected Altithermal), are all consequences of Gifford Pinchot's poor decision to adopt fire suppression for America in 1905. Pinchot however, is by no means the only one at fault. The political party system itself is the true culprit because it has allowed politicians to elevate pseudointellectuals into positions of prominence and influence within the bureaucracy. These overly-powerful individuals went on to make poor decisions well beyond their own realm of understanding. The consequences were made worse because mistakes were often played down or hidden from the

American people.

I firmly believe many of these past poor decisions were made with the best of intentions, but not all. Over 30 years ago, the passions generated by the environmental movement gave rise to a religious like ideology derived when the incompatible Murr and Pinchot doctrines combined, which seized control of the American Democratic Party. The Democratic Party died and went forward in name only. The new age political ruling class installed political religionists into leadership positions within the powerful bureaucracy.

Ideologically driven, these individuals soon not only occupied positions of power throughout the federal government, but also dominated academia, the judiciary, the mainstream media and Hollywood. They saw themselves on a quest to save the world while frequently living lives of privilege, isolated from the negative consequences created by the industrial revolution. This insolated existence has caused most of them to remain blissfully unaware of their own personal role in usurping the founding principles of America. Their actions have allowed an all powerful Marxist like political ruling class to slowly take control of America. This tenure of centralized control has allowed environmental problems to fester, which are now teetering near the breaking point. The clergy (lawyers, politicians and professors) of this political religion have removed and/or destroyed the only people capable of innovating our way out of this mess.

The political ruling class is now using the COVID-19 pandemic and any other moment of crisis as a means of destroying the economy, in hopes of finishing off America's small business owners. They are doing so in order to dismantle the risk-taking middle class, which up until now has been the only thing preventing these Marxist elitists from centralizing all authority onto themselves.

'We the People' must quickly take back control from the political ruling class by abolishing the political party system, which is the source of their power. The time has come to harness the enormous empathy and intellectual might of America by placing the power as was intended, in the hands of the American people. We can accomplish this by amending the Constitution to include the 4th branch of government, as described towards the end of Chapter 1.

To ignore our current situation, is to usher in the Biblical Armageddon.

The future is not yet set. It will form from the 'decisions' we make today. We as individuals must choose correctly and step away from our political tribes.

Contact your elected officials and demand the political party system be dissolved and replaced with the 4th branch of government. Those politicians that resist will have just revealed themselves as members of the political ruling class.

THE BREAKTHROUGH

In 2003, I knew I'd succeeded in creating an industry that was capable of breaking the ultra small tree barrier. I immediately went over to Dad and Mom's house to present this latest iteration. There'd been many such meetings over the course of the previous ten years or so.

As I've stated earlier, Dad had changed after the 'Earth First' attack. His formerly open mind had slammed shut on matters concerning the environmental movement. He now saw them, and justifiably so, as anti-American terrorists/Marxists.

You may have noticed that my theories, while not brutal, can be taken as less than kind in regards to the timber industry. Dad thought I sounded a little too much like an environmentalist. Other loggers had thought so as well, so I'd stopped talking about my ideas to almost everyone except Dad.

Throughout the preceding years, as I struggled with how to devise the desperately needed industry, I would go over and present my latest concept to Dad. He would immediately find the flaw and exclaim his discovery with smiling satisfaction. In his mind he was protecting European style tree farming and although he never said it, he was getting a little revenge against a movement that had taken so much from him. Needless to say, his newfound strong bias and high intelligence made him the perfect candidate to bounce ideas off of.

Sometimes months would pass before I brought him my next concoction. The result however, was always the same; he would find the hidden flaw. As the years passed, there were many such meetings with each ending the same way. It was however, becoming harder and harder for him to find the unworkable issue.

The meeting in 2003 was different. I knew I had it. After arriving, I described my process. He mulled it over and over and couldn't find a problem. The night eventually ended with many questions and answers exchanged, yet he remained unable to find the unworkable issue. It was about a week later when I saw him again. He told me that he'd thought of little else. He then gave me an answer.

It was an answer that still haunts me to this day. Dad said, "They will never let you do it", meaning the politician ideologues that were masquerading as environmentalists. To bolster his statement, he then pointed to the egregious injustice that had taken place in the early 90s when we loggers had watched helplessly as the veteran federal foresters were forced out and replaced with party loyalists. The ideological takeover had only picked up speed from that moment forward. The political ideologues were now in control of every aspect of power and influence throughout our society.

Dad's sympathy was genuine when he said "Sorry", which meant he believed I had the solution, but that I'd wasted my time.

Undeterred, I set up a meeting with an official from the State of Montana. It was during that embarrassing encounter that I quickly realized I couldn't adequately explain why we needed this industry. I'd failed because although I'd been looking, I hadn't been able to find documentation which proved conclusively that the Native American burning had been both widespread and beneficial. This new obstacle seemed insurmountable. After a couple years, I went around that barrier when I found Stephen Arno's book, 'Northwest Trees'.

I studied it intently for years, eventually identifying the two different tree groups, Inter-glacial and Glacial. This was ground breaking; no one had ever put that together. I soon became aware of the mass extinction event and the mastodon. I could now explain how the Native American's succeeded. In 2009, I began writing a book.

It was shortly thereafter that I found Omer Stewart's book, 'Forgotten Fires', which documents pre-settlement Native American burning traditions. It was everything I'd been searching for. Within its pages are conclusive evidence of not only widespread application, but the why and how.

The two introductions for 'Forgotten Fires' are however a little disturbing and written by two other scientists. They spoke of the reasons why Omer had been unable to find a publisher during his lifetime.

He'd written his book in the early 1950s. One reason he wasn't able to publish was simply bad timing. The Pinchot doctrine had just made fire suppression wildly popular with the 'Smokey the Bear' ad campaign. The other reason has everything to do with twelve federal court cases, in which Omer had testified on behalf of the Native Americans. His expert testimony had been instrumental in obtaining a positive verdict in each case. In so doing, Omer had made an enemy out of the politically-powerful Julian Steward, who was supposedly the nation's leading anthropologist at the time. Julian also happened to be Omer's former professor. Julian had defended the government position during the previous litigation, in a failed attempt to reduce the size of several western Native American reservations.

In the end, Omer had made his opponent look rather small and unprepared. Sour grapes ensued, Omer became blacklisted and soon found himself unable to find employment within a university possessing a strong anthropological department. He attempted to publish his book several times. Each attempt proved unsuccessful. Omer passed away in 1993, his manuscript forgotten and lost.

Eventually, it was rediscovered and published. Between its covers, Omer brings forth among a plethora of other important information, an original theory. He believed an ancient people at some point in the distant past removed the Great Plains forest.

I immediately stopped writing and dove in deep. The next few years were interspersed with epiphany after epiphany. By 2016, I had everything, which included the true reason as to why the climate has begun oscillating. My own theory had now grown well beyond simply developing and installing a new type of forest industry, presenting me with same old problem: who do I approach?

For reasons that should now be obvious, I couldn't go to the mainstream media, academia or anyone within the government. There was however, one bright option. Facebook would allow me to speak directly to the American people or so I then thought. In the intervening five years I've learned plenty. Such as, if you don't belong to a political faction, you don't exist. I also discovered that Facebook is operated by political ideologies.

In the absence of divisive politics, social media is our way out of this mess. I came to this realization during the winter of 2019-2020 and set about devising the basic concept for the 4th branch of government. Yet Dad's long ago response continues to play over and over in my head. "They will never let you do it." This always leads to another still unanswered question: "Is this very real solution going to die with me?" That question can only be answered with the passage of the 4th branch of government.

In the meantime, join me on the social site MeWe. Ask questions and make suggestions on how the 4th branch should be structured, or simply indicate which ideas appeal to you. I will use the proceeds from this book to begin building the wide array of equipment that this still-imaginary industry requires. When fully realized, this industry will be able to resurrect the proper climate-stabilizing forest structure, which will finally allow the old growth tree species to live out their entire lifespans.

At which point, I will truly be 'The Last Old Growth Sawyer'.

The End... maybe.

This photo was taken sometime in 2013. I'm flanked by my hardhat wearing nephew Josh Bruch (chocker setter) and my step-son Chris Stein (Yarder operator and a U.S.M.C. Afghanistan war veteran). Photo by Jamie Pein

ONE MORE CHAPTER
FIGHTER-ATTACK SQUADRON - VMFA 112 AND BATTLING POLITICAL PYRAMID SCHEMES

I would like to tell one last story and mention a few, who served with me in Marine Corp fighter/attack Squadron VMFA 112 'Cowboys'. These Marines either worked on or flew the mechanically complex and maintenance demanding F-4 Phantom II.

F-4 Phantom II from U.S. Marine Corps squadron VMFA 112

The circumstances surrounding the above photo remain vivid. It was the final phase of the Cold War. We were practicing hard for a potential conflict with Communist Soviet Union, which thankfully never materialized.

This Phantom was one of two taking off in the twilight from the temporary steel runway on Marine Corps base Twenty Nine Palms during CATX (combined arms training exercise '84). Each Plane loaded with 6 - two thousand lbs. bombs.

Some time later, the ground crew watched this Phantom (flown by legendary fighter/attack pilot Colonel Mike 'Lancer' Sullivan and RIO'd by Major Bill 'B-Bop' Bower), along with the other F-4, descend in a steep dive on the distant mountains. It was fairly dark now, which made the flares they were popping (counter measure for heat seeking missiles) look all the more dramatic. Our silent tension quickly giving way to desperate mutterings of 'pull up' as it became apparent that they'd flown too low and wouldn't be able to recover... Just then… they released their bomb load and pulled up hard. The explosions were visible and soon audible to their ground crew audience. We yelled and cheered our approval at this extreme 4th of July like exhibition.

Unbeknownst to us was that a bomb rack had malfunctioned, causing a bomb to hang up briefly before falling. The bomb's tail fin puncturing about a 4" hole under the right wing of Sullivan's F-4. Fortunately, the aircraft landed safely.

As a Plane Captain it was my job to launch, recover, fuel, inspect and make the aircraft ready for flight.

The aircrew crawled out of the Phantom in the now pitch black darkness. We had a brief discussion about the previous sortie. Once completed, they turned around with flashlights on and walked to the tent, which was serving as maintenance control. I began readying the aircraft for the next flight. I soon found the hole. I immediately reported my find to maintenance control (SSGT Gary Morris). A short time later, Metal shop was looking up at the gapping hole (Phantom Squadrons possessed about a dozen specialized repair shops, Hydraulics, Com/Nav, Powerplants, etc.). It was soon determined that the puncture was located in an unfavorable location, which may cause the aircraft to be sent back to the factory for repair.

Col. Sullivan soon learned of the maintenance debate. He walked back out on the flight line, stiffly bent over and walked under the wing. He shined his flashlight up into the unwanted void. His considerable stiffness perhaps attributed to his recent flight, but more likely caused by the cumulative effect of flying the Phantom for the previous two decades. It is rumored that he has acquired more flight hours in an F-4 Phantom II than anyone else in history.

He glanced up for a moment and backed out from under the wing. Col. Sullivan's highly competent leadership style garnered him the genuine respect/loyalty from all who served with him, which (at least to me) also seemed to somewhat resemble the onscreen persona of John Wayne. He turned to Morris and said, "We flew worse then that in Vietnam, put a patch on it".

It took the remainder of the night, but Metal shop succeeded in patching the hole. The aircraft flew the next morning. The patch held, until sadly, this and almost all other Phantoms were deemed obsolete and sent to the boneyard a few years later. Many to be used as target practice.

FIGHTING COMMUNISM

During the Cold War, the United States military leadership studied soviet military tactics intensity. Those of us serving, hoped that knowledge would give us an advantage in what seemed to be an inevitable armed conflict.

Then Communist Soviet Union collapsed... Problem solved...right...

When I witnessed the behavior of those angry and unreasonable environmentalists at Challenge creek in 1992. I viewed it as an undeniable indicator that the environmental movement had died and turned into something sinister.

For years, I thought it was a Communist attempt to take over America, but that isn't what happened, nor is that what occurred during all those decades in which 'Communist' Russia existed. The problem began long ago and is far simpler.

The industrial revolution has ushered in many wonderful scientific advances, but it came at a steep cost to individual freedoms. People the World over suddenly found themselves increasingly subservient to powerful centralized authorities, both monetarily and governmental.

In an effort to solve this very real problem, a few Russians overthrew Czar Nicholas II and installed Communism or at least that was their intention. In reality, Lenin and Stalin only succeeded in centralizing all power on to themselves. Individual freedoms died along with of ten's of millions of Russians.

Hitler's socialist approach seemed different, but the end result was exactly the same. Millions died with all Germans forced to answer to an all powerful Hitler and his inner circle.

Zedong Mao installed Communism in China, which resulted in the starvation death of millions of his own people (I'm sure Mao didn't miss a meal). This horrific occurrence gave rise to a phrase in America during the 1950s and 1960s, with Parents telling their children to 'eat everything on their plate', because children in China were starving. The list of these murderous regimes goes on and on. They promised utopia, but only delivered various types of hell on Earth.

The true reason, that over 100,000,000 people died during the last century, was the direct result of the rise of all powerful 'Political Pyramid Schemes'. Usurpers overthrew the previous government by appealing to an almost universal desire for equality and/or scientific progressivism. The successful revolutionaries were equally enthusiastic, but the intellectually limited centralized authority quickly found itself incapable of delivering on what it had promised. So they began imprisoning or killing anyone who disagreed with their leadership style. The power structure formed into a pyramid as the politically agreeable are given high positions of power and privilege within the bureaucracy. The vast majority of the population soon found themselves powerless and residing at the bottom of the pyramid, which constricted diversity of thought and stifled innovation.

Truth was a constant threat to those residing at the top. So in order to suppress truth, the State run media and the education systems were harnessed to both propagandize and indoctrinate the citizenry.

In the Soviet Union, the gulag system was created for the 'bottom of society', which included anyone who disagreed with the all powerful ruling class (a book written by Aleksandr Solzhenitsyn, 'The Gulag Archipelago' provides a horrific account). Germany had its own version. China still does.

This same type of 'Political Pyramid Scheme' had already gained traction within America by 1992, which was evidenced by the arrival of that angry van load of environmentalists at Challenge creek. It was heartbreakingly apparent, that these well intended people no longer sought genuine discussions/solutions.

Instead, they became part of the rising culture of complaint, which seemed only interested in assigning blame onto other groups. Groups that just also happened to be the backbone of America's economy (these outed groups included several American occupations, which helped feed, cloth, house and power the World). Remarkably, these closed minded complainers did this while remaining blissfully unaware of their own role in the culvert political scheme, which logically, could only result in the eventual rise of an all powerful centralized authority.

Smear campaigns produced by a complicit main stream media, targeted Loggers, ranchers and many others living in rural America. These innocent groups slid to the bottom of the pyramid. I marveled at just how fast we loggers had been character assassinated.

The destruction of the private sector economy is pursued, not out of environmentalism, but to ensure that the government will eventually be the only option available for societal advancement. Once complete, this tactic effectively places everyone under the control of those at the top of the power pyramid.

America's education system slowly ebbed into indoctrination centers.

The U.S. Federal Bureaucracy was rapidly expanded (support among federal employees for one particular political party eventually reaches 90%, while the other is rendered meaningless. Today, the federal 'public servant' is financially compensated at an average rate two and a half times higher, than that of the tax paying private sector employee). Millions of private sector jobs are exported to Communist China. The conditions in that country are so soul crushing, that suicide nets are placed around high rise factories.

Here in America, a political campaign of hate and division only pick up speed.

The most recent group to be targeted for social assassination are the Police. This isn't by accident. The political ruling class is hoping that the honest and compassionate Cop will quit. So they can be replaced with a complaint party loyalist.

America's all consuming/powerful political pyramid is now complete.

The World has been down this well trodden road several times. This time however, it leads not only to human suffering, but climate/ ecological collapse and the lake of fire.

The proposed 4th branch of Government will save us from this fate, by decentralizing thought, which will unleash innovation by creating a hierarchy of competence. Perhaps, even ushering in a genuine Utopia.

Semper fi to my Cold War brothers and Sisters
Alejandro Montalvo, Joe Crowdus, Andy Arndt, Lisa DiDonato, John Cole, John Uchniat, Cliff Prouty, Ron Newhart, Scott Rasmussen, Curt Holle, Rick Meredith, Tony Gonzales, Dwain Elkins (U.S. Navy veteran), Bob Kinniburgh (U.S. Army Vietnam War veteran) and all those many many others not mentioned.

I would also like to remember three of VMFA's 112 Phantom pilot's, who passed away in two separate flight mishaps in the last half of the 1980s. Capt. Rob 'Pork' Tucker (a ground crew favorite), Capt. J Bourne (flying back seat with Tucker) and Major Jim 'Greyhound' Simons (The only other Squadron member from Montana).

The End… or possibly…
this is only the beginning.

In the end, Bruch brothers line logging consisted of Just Shawn and myself.

There have been many others over the years who have worked with Shawn and I, but haven't been mentioned in previous pages. They however were equally indispensable in helping Shawn and I stay in the forest and therefore enabled discovery.

Here is a partial list: Bob Wagar (legendary log truck driver), Chans Wyant (nephew/chocker setter), Skyler Bishop (nephew/processor operator), Dustin Eklund (nephew/Chocker setter), Carla Bruch-Breland (Sister/Chocker setter), Dan Milliron (chocker setter), Joe Bybee (legendary mech logger. He took a chance on Shawn and I, when few others would and financed Bruch Brothers line logging's first Timbco in 2001), Uncle Matt and Aunt Cindy Eklund (who without hesitation, lent us the down payment for that first Timbco), Les Dennison (Aunt Virginia's son/Master Machinist), Adam Tisdale (Master Mechanic and U.S.M.C. veteran) and Brad Gorton (Legendary master mechanic). Will Hankinson (Expert heavy equipment transporter), Chad Marquardt (chocker setter), Darwin Marquardt (yarder operator and one of the last homestead descended ranchers, in NW Montana), Jerry Schlegel (Expert log hauler), Northwest Parts and Equipment.

Olde Stonewall Golf Course near Ellwood, PA.

I would also like to give a special thanks to members of my writing group: Craig Naylor, Effie Roma, Jami Smith and Dan McGoldrick (U.S. Army Iraq war veteran). Who, once a week for six months were both my writing critics and sounding board. They were the first to be subjected to the depths of my theory. From these interactions spawned my realization that the 4th branch of Government needed to be devised and implemented. The initial spark for this interconnected advisory platform was ignited from a five minute conversation I had with the President of 'The authors of the Flathead', Betty Kuffel MD. In which she described the book she'd written 'Fatal Feast' that discusses the prions pandemic (Chronic Wasting Disease…CWD) which is sweeping wildlife populations around the globe. I quickly realized the spread of this pathogen is also a product of fire suppression and my proposed widespread Inter-glacial forest restoration program will extinguish this issue. This is something I might have never realized, if we'd never met. The 4th branch will bridge the practical world with the scientific and usher in and establish an age of generational continuity. This is imperative, because it will take several generations of unity and peace to achieve success ... Thank you all.

I would also like to thank all my many other 'professors' not listed. Together, we've unlocked these very real World Saving Solutions...

The following list of references are intentionally incomplete. My theory is merely a starting point. I encourage you to dive in deep and uncover the truth. Truth does not naturally reside in the human heart. It can however be found in nature.

Forgotten Fires by Omer C. Stewart
Northwest Trees by Stephen F. Arno
Plow A New Furrow by Emma Hedman
Year of the Fire by Stephen J. Pyne
Fatal Feast by Betty Kuffel
Beyond The Hundredth Meridian by Wallace Stegner
The Wilderness World of John Muir, edited by Edwin Way Teale
The Three Hundred Year War, A Chronicle of Ecological Disaster by William O Douglas
Frozen Earth by Doug Macdougall
Nature and Other Essays by Ralph Waldo Emerson
Glacial Lake Missoula by David Alt
Montana Before History by Douglas H. Macdonald
Twilight of the Mammoths by Paul S. Martin
Pleistocene Extinctions by Paul S. Martin and John E. Guilday
Undaunted Courage by Stephen E. Ambrose
The Bible
The Quran
The Forests of Northwest Montana

INTERNET SEARCH WORDS AND PHASES
Pleistocene
Mega Fauna extinction event
Mastodon
Mammoth
Black Mat
Ice Core analysis graph
Holocene
Clovis culture
Clovis points
Micro charcoal
And on and on and on…

Made in United States
Orlando, FL
10 November 2026

90498870R00069